Math Competencies
for Everyday Living

Jeanette Powell
Olympic High School
Concord, California

Barbara Hartley Scott
Olympic High School
Concord, California

Teacher Consultant:
Claudia Carter
Mississippi School for Mathematics and Science
Columbus, Mississippi

MO7
PUBLISHED BY
SOUTH-WESTERN PUBLISHING CO.
CINCINNATI   WEST CHICAGO, IL   CARROLLTON, TX   LIVERMORE, CA

Copyright © 1990
by South-Western Publishing Co.
Cincinnati, Ohio

ALL RIGHTS RESERVED

The text of this publication, or any part thereof, may not be reproduced
or transmitted in any form or by any means, electronic or mechanical,
including photocopying, recording, storage in an information retrieval system
or otherwise, without the prior written permission of the publisher.

ISBN: 0-538-13072-5

Library of Congress Catalog Card Number: 88-64145

2 3 4 5 6 H 4 3 2 1 0

Printed in the United States of America

CONTENTS

PREFACE TO STUDENTS — v

PREFACE TO TEACHERS — vii

UNIT 1 THE CALCULATOR AND A REVIEW OF THE BASICS

Using the Calculator	2
Determining the Place Value of Whole Numbers and Decimals	4
Rounding Numbers	9
Decimals	13
Fractions	20
Equivalents: Fractions = Decimals = Percents	36
Ratio and Proportion	41
Percent	46
Other Approaches to Solving Percent Problems	50

UNIT 2 WORD PROBLEMS AND PROBLEM SOLVING

Strategy 1—Follow Directions	60
Strategy 2—Read Problem Carefully and Look for Key Words	64
Strategy 3—Decide Which Math Operation to Use	68
Strategy 4—Estimate	70
Strategy 5—Simplify the Problem and Substitute Easier Numbers	74
Strategy 6—Look for Likenesses and Patterns	80
Strategy 7—Guess and Check	85
Strategy 8—Draw a Picture or Diagram	87
Strategy 9—Manipulate (Touch and Move) Objects	89
Strategy 10—Keep Track of Clues and Information	92

UNIT 3 COMPARISON SHOPPING

Shopping for Groceries	100
Miscellaneous Purchases	113
Automobile Expenses	116
Cost of Credit	118

UNIT 4 TABLES, GRAPHS, AND AVERAGES

Reading Tables	124
Reading Graphs	126
Making Graphs from Given Information	137
Practice with Tables and Graphs	144
Finding Averages: Mean, Median, Mode	148

UNIT 5 MEASUREMENT

Linear Measurement	154
Metric Measurement	163
Cooking Measurement	171
Temperature Measurement	175

UNIT 6 BANKING

Opening a Bank Account	180
Deposit Slips and Checks	181
Filling Out a Check Register	190
Bank Statements	192
Automatic Teller Machines (ATM)	194
Checking Your Balance	196

UNIT 7 GEOMETRIC SHAPES AND CALCULATIONS

Geometric Figures	200
Perimeter	204
Area	211
Volume	218

UNIT 8 TIME

Figuring Time Intervals	222
Time at a Job	229
Using Transportation Schedules	233
Using Time Zones	235

INDEX 237

PREFACE TO STUDENTS

Have you ever wondered why you need math and where you will use math skills in your life? *Math Competencies for Everyday Living* teaches basic math skills with the use of the calculator. It has problems that use math in situations such as shopping, banking, cooking, and interpreting information in tables and graphs.

FEATURES

Math Competencies for Everyday Living is a text-workbook with the following features.

Simple to Complex

Math Competencies for Everyday Living is arranged so that easy problems are taught and practiced before you move to more difficult problems of the same nature.

Examples

Examples are presented whenever a new concept is taught.

Challenging Problems

Problems in this text-workbook vary in difficulty. Some are easy and some are very challenging and will test your problem-solving abilities. It is important to continue working on a challenging problem even though you do not solve it on your first try. Continue using a variety of problem-solving strategies until you solve the problem.

Group or Individualized Instruction

Math Competencies for Everyday Living is designed to allow you to work independently or in a group or in a classroom situation.

Student Audience

Math Competencies for Everyday Living is written for the student who wants
- a review of basic math skills
- practice for a high school math competency test
- applications of basic math skills to situations found in everyday living
- practice in problem solving.

PLAN OF PRESENTATION

There are eight units in this text-workbook. Each unit is introduced by a page that lists what skills are needed to start the unit, what skills you will learn, and the situations in which you will use these skills.

The eight units in *Math Competencies for Everyday Living* are listed below.

Unit 1   The Calculator and a Review of the Basics
          This unit makes up the first 25% of the text. Place value, rounding, decimals, fractions, equivalents, ratio and proportion, and percent are reviewed in Unit 1.
          This review is introduced through the use of the calculator, and your use of the calculator is encouraged throughout the entire text-workbook.

Unit 2   Word Problems and Problem Solving
          Unit 2 is a unique presentation of problem solving. Ten problem-solving strategies are identified, explained, and modeled. Practice for each of these problem-solving strategies is provided.

Unit 3   Comparison Shopping

Unit 4   Tables, Graphs, and Averages

Unit 5   Measurement
          This unit includes linear, metric, cooking, and temperature measurement.

Unit 6   Banking

Unit 7   Geometric Shapes and Calculations

Unit 8   Time

PREFACE TO TEACHERS

Math Competencies for Everyday Living shows students the value of learning math skills, teaches students problem-solving techniques, fosters a positive attitude in students towards math, and encourages students to use the calculator to review basic math skills.

FEATURES

Math Competencies for Everyday Living is a text-workbook with the following features.

Simple to Complex

Math Competencies for Everyday Living is arranged so that easy problems are taught and practiced before students move to more difficult problems of the same nature.

Examples

Examples are presented whenever a new concept is taught.

Challenging Problems

Problems in this text-workbook vary in difficulty. Some are easy and some are very challenging and will test students' problem-solving abilities. It is important that teachers model approaches for solving the challenging problems and that students continue working on the problems even though they may not solve them on their first try.

Group or Individualized Instruction

Math Competencies for Everyday Living is designed to allow students to work independently or in groups or in a classroom situation. It can be used as a text for a basic math class, for a remedial math class, or as a supplemental text in classes that teach consumer living skills.

Student Audience

Math Competencies for Everyday Living is written for the student who wants

- a review of basic math skills
- practice for a high school math competency test
- applications of basic math skills to situations found in everyday living
- practice in problem solving.

PLAN OF PRESENTATION

There are eight units in this text-workbook. Each unit is introduced by a page that lists what skills are needed to start the unit, what skills the students will learn, and the situations in which students will use these skills.

The eight units in *Math Competencies for Everyday Living* are listed below.

Unit 1   The Calculator and a Review of the Basics

This unit makes up the first 25% of the text. Place value, rounding, decimals, fractions, equivalents, ratio and proportion, and percent are reviewed in Unit 1.

This review is introduced through the use of the calculator, and the student's use of the calculator is encouraged throughout the entire text-workbook. Using the calculator makes the review more enjoyable, and gives the student practice in using a tool commonly used in the adult world.

It is recommended that the $\frac{\%}{100} = \frac{Part}{Total}$ method be used to teach percent to all students who have not already learned percent by other approaches.

Unit 2   Word Problems and Problem Solving

Unit 2 is a unique presentation of problem solving. Ten problem-solving strategies are identified, explained, and modeled. Practice for each of these problem-solving strategies is provided.

Many students have basic math skills and yet have difficulty with word problems and problem solving. It is recommended that teachers model various approaches to solving problems. It is especially important that teachers model the processes to solve Strategy 7 (Guess and Check), Strategy 8 (Draw a Picture or Diagram), and Strategy 10 (Keep Track of Clues and Information). These strategies may be unfamiliar to students, and it would be very helpful for them to see some of the problems worked.

The review test for Unit 2 requires that the student use and identify each of the ten strategies as the problems are solved.

Unit 3   Comparison Shopping

Unit 4   Tables, Graphs, and Averages

Unit 5   Measurement

This unit includes linear, metric, cooking, and temperature measurement.

Unit 6   Banking

Unit 7   Geometric Shapes and Calculations

Unit 8   Time

PHILOSOPHY

One of the main goals of *Math Competencies for Everyday Living* is to provide opportunities for students to apply their math skills in everyday situations. Another important goal is to help students become problem solvers. In order to become successful problem solvers, students need to understand and accept that they may be perplexed and unable to solve some problems the first time they try. It is strongly recommended that recognition be given to a student who tries several different strategies to find a solution to a problem. The "I can" and "I will continue trying" attitudes are important. These attitudes are more important to problem solving than getting the correct answer on the first try.

SOLUTIONS

Some of the problems in the text can be solved in more than one way. Encourage students to seek alternative methods to solve problems. Some of the problems ask for approximations, allowing students to use estimating skills.

TEST PACKET

A test packet accompanies this text-workbook. Included are the following items.

- A survey test that has been divided into two parts so that students may complete a part in one period and not feel overwhelmed by the length of the survey. For traditional math classes the survey test may be used to group students by areas of need. In individualized classrooms the survey results indicate which skills a student needs to review.
- A review test for each of the eight units.
- A Student Survey and Record Sheet on which daily work and review tests can be recorded.
- Manipulative pages for fractions and measurement.

1
THE CALCULATOR AND A REVIEW OF THE BASICS

SKILLS YOU WILL NEED

Reading
Following directions

SKILLS YOU WILL LEARN

Using a calculator
Understanding place value
Rounding
Working with decimals
Working with fractions
Determining equivalents
Understanding ratio and proportion
Working with percents

SITUATIONS IN WHICH YOU'LL USE THESE SKILLS

At home
On the job
In the marketplace

The calculator is a useful tool for quickly and accurately working everyday math problems. Different models of calculators may differ slightly. For example, some calculators display the total before the equals key is pressed. However, they all operate in much the same way. You may need to read the directions for your calculator and experiment with it while working on the exercises that follow.

Definitions of Keys

AC	Clear
CE/C	Clear/Entry
.	Decimal Point
÷	Divided by
=	Equals
−	Minus
%	Percent
+	Plus
×	Times

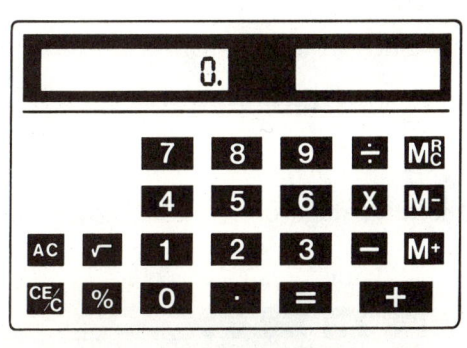

HINT: The calculator shown here is solar powered. Therefore, it does not have the usual ON/OFF key.

UNIT 1 ■ THE CALCULATOR AND A REVIEW OF THE BASICS   1

USING THE CALCULATOR

Enter the following numbers and symbols into your calculator and answer the questions. Answers for these exercises may vary according to your calculator's design and capacity. Learning what *your* calculator can do is more important than getting the "right" answer.

Clearing All Entries and Clearing Only the Last Entry

1. [AC] 1 [+] 1 [+] 1 [=] _____

   Can you completely clear *your* calculator by pressing [AC] once? _____

2. [AC] 1 [+] 5 [+] 3 [CE/c] [=] _____

   What happened to the 3? _____

3. [AC] 5 [+] 3 [+] 8 [AC] [CE/c] _____

   Which key clears all entries on your calculator? _____

4. [AC] 13 [+] 7 [+] 3 [CE/c] [=] _____

   Which key clears just your error? _____

Multiplying with the Equals Key

> **HINT:** Multiplying with the equals key does not work on all calculators.

1. [AC] 9 [×] 9 [×] 9 [=] _____

2. [AC] 9 [×] [=] [=] _____

   What number did you see when you pressed the first [=]? _____
   What number did you see when you pressed the second [=]? _____

3. How many times would you press the equals key if you wanted to multiply 13 [×] 13 [×] 13 [×] 13 [×] 13? _____

   What is your answer? _____

4. What is the advantage of using the equals key for multiplying? _____

5. Press the calculator keys as directed.
   A. [AC] 2 [=] [=] [=] [=] _____
   B. [AC] 2 [×] [=] [=] [=] [=] _____

6. Press one number key repeatedly until your screen is full. How many digits are displayed? _____

Using Other Function Keys

M+   Puts a number in the memory of the calculator.

M−   Removes the numbers from the calculator's memory.

MR/C   A two-function key. If you press it once, the key becomes **MR** and *recalls* (brings back from memory). If you press it twice, the key becomes **MC** and *clears* the calculator's memory.

Practice using the memory keys on the following problems.

1. AC 2 + 3 = M+ 4 + 3 = M+ MR _____
2. AC 5 + 8 = M+ 6 + 7 = M+ MR _____
3. AC 1 + 2 = M+ 2 + 3 = M− MR _____
4. Now press **MR/C** again. What shows on the display? _____
   What did you clear by pressing **MR/C** twice? _____
5. AC 8 − 5 = M+ 12 − 9 = M+ MR _____
6. AC 5 + 3 = M+ 2 + 3 = M+ M− 5 MR _____
7. AC 15 + 18 = M+ 35 + 56 = M+ MR _____
8. CE/C 10 − 2 = M+ 6 − 2 = M+ MR _____
9. Why wasn't your answer 12? _____
10. AC 3 + 5 = M+ 5 + 6 = M+ 8 − 3 = M+ MR _____
11. AC 25 + 3 = M+ 18 + 5 = M+ M− 23 MR _____
12. In Problem 11, the second **M+** entry was an error. How can an error be removed from the calculator's memory? _____
13. Hector was shopping for food for a beach party. He had $25 to spend. He purchased 1 bottle of catsup for 80¢, 3 packages of hamburger buns for $1.09 each, 5 pounds of hamburger at $1.70 a pound, 4 cartons of juice at 95¢ each, and 3 packages of nuts at $1.65 each. How much money will he have left after shopping? _____

UNIT 1 ■ THE CALCULATOR AND A REVIEW OF THE BASICS

DETERMINING THE PLACE VALUE OF WHOLE NUMBERS AND DECIMALS

The place a digit occupies in a number gives it its value. A number to the right of a decimal point is less than a whole number.

Place Value Chart

The chart below illustrates the place values for the most commonly used whole numbers and decimals.

Whole Numbers							and	Decimals		
millions	hundred thousands	ten thousands	thousands	hundreds	tens	ones	decimal point	tenths	hundredths	thousandths
9	2	0	8	3	6	5	.	7	4	1

In the number 35,789.164, what is the place value of each of the following digits?

1. 6 _____
2. 7 _____
3. 4 _____
4. 9 _____

5. 5 _____
6. 1 _____
7. 8 _____
8. 3 _____

Change the following words to numbers. (Remember to put in the decimal point.)

1. five tenths _____
2. sixty-one hundredths _____
3. seventy-three thousandths _____
4. one tenth _____
5. three hundredths _____
6. nine tenths _____

UNIT 1 ■ THE CALCULATOR AND A REVIEW OF THE BASICS

When a number is written in words, the word *and* indicates the decimal point. Change the following words to numbers.

1. Two and eight tenths — 2.8
2. Sixteen
3. Fifteen hundredths
4. One and five hundredths
5. Seventy-five hundredths
6. Five and five hundredths
7. Ten and sixty-five hundredths
8. Eight thousandths
9. Seven and seven tenths
10. Six and fifty-two hundredths
11. Twenty-five hundredths
12. Three and three thousandths
13. One tenth
14. Sixty-three and sixty-two thousandths
15. Four and sixty-eight hundredths
16. Nine tenths
17. Two hundred twenty-eight and thirty-one hundredths
18. Six and eight tenths
19. Seventy-eight
20. One hundred fifty-six and five tenths
21. Four and six hundredths
22. One hundred thirty-seven and three tenths
23. Twenty-two and thirty-seven hundredths
24. Three and four tenths

Fill in the blank spaces below with the correct words or numbers. (Remember to clear your calculator by pressing **AC** before each number.)

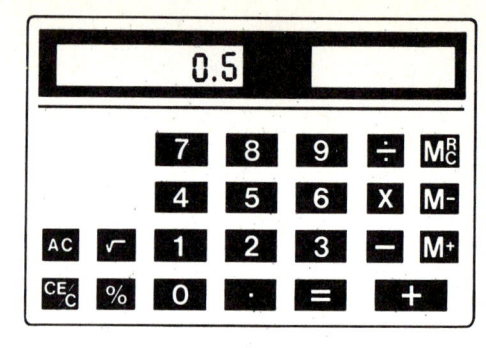

Write It **Key It**

1. five tenths — 0.5
2. two and eight hundredths — _____
3. _____ — 0.15
4. five and five thousandths — _____
5. nine hundredths — _____
6. _____ — 63.63
7. _____ — 27.7
8. five and three tenths — _____
9. two and four hundredths — _____
10. _____ — 248.2
11. fifty-five hundredths — _____
12. _____ — 60.039
13. _____ — 49.48
14. eight and five tenths — _____
15. _____ — 400.16
16. one and seven hundredths — _____
17. _____ — 25.4
18. _____ — 10.28
19. two and two hundredths — _____
20. two and two thousandths — _____

6   UNIT 1 ■ THE CALCULATOR AND A REVIEW OF THE BASICS

Visual Explanation of Decimal Place Value

Look at the square to the right. Note that one tenth (0.1, $\frac{1}{10}$) is the largest darkened area. One hundredth (0.01, $\frac{1}{100}$) is *smaller* than one tenth. One thousandth (0.001, $\frac{1}{1,000}$) is *smaller* than one hundredth.

$$\frac{1}{10} = 0.1$$

$$\frac{1}{100} = 0.01$$

$$\frac{1}{1,000} = 0.001$$

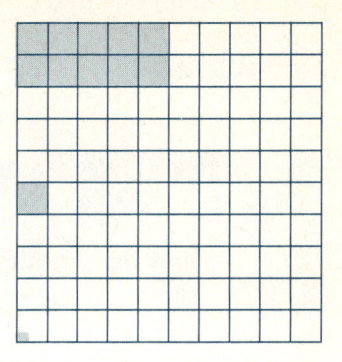

Shade and label the squares on the right as directed.

1. Square A:
   0.3
   0.03
   0.003 (approximately)

Square A

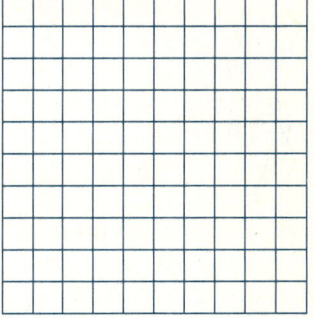

HINT: Shaded squares may vary in arrangement for numbers 1 and 2.

2. Square B:
   0.5
   0.05
   0.005 (approximately)

Square B

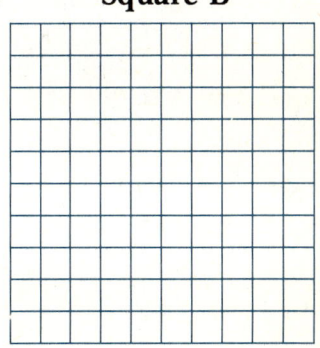

Circle the *smaller* number.

1. 0.3 or 0.03

2. 0.04 or 0.004

3. 0.023 or 0.23

Circle the *larger* number.

1. 0.009 or 0.9

2. 0.2 or 0.02

3. 0.19 or 0.3

UNIT 1 ■ THE CALCULATOR AND A REVIEW OF THE BASICS

A value of less than one can be written as a decimal or a fraction, or it can be combined with a whole number. Study the examples and change the following words to decimals and fractions.

	Words	Fractions	Decimals
1.	three and seven tenths	$3\frac{7}{10}$	3.7
2.	four and one hundredth	$4\frac{1}{100}$	4.01
3.	one thousandth		
4.	two and two hundredths		
5.	four hundredths		
6.	four tenths		
7.	four hundred four and forty thousandths		
8.	eight and six hundredths		
9.	fourteen thousandths		
10.	twenty-six hundredths		
11.	eighteen and six tenths		
12.	fifty-seven and five hundredths		
13.	twenty-six thousandths		
14.	six tenths		
15.	four and eight hundredths		
16.	fifty-nine and five thousandths		
17.	six hundred twenty-three and two tenths		
18.	forty-seven and thirteen hundredths		

Circle the numbers in each row that have a 5 in the *tenths* place. Add those numbers with your calculator and write your answer in the blank.

Example   53.35; (7.5); 25.4; 68.05; (2.51); 51.6   _____10.01_____

1. 685; 2.57; 0.451; 35.57; 6.25; 5.05   _____
2. 51.6; 75.2; 345; 0.567; 127.5; 0.275   _____
3. 7.75; 8.54; 552.255; 2.59; 8.75; 7.56   _____

Circle the numbers in each row that have a 7 in the *hundredths* place. Add those numbers with your calculator and write your answer in the blank.

1. 717.86; 4.87; 87.78; 2.7; 4.276; 9,751   _____
2. 0.297; 758; 9.876; 20.71; 8.678; 3.67   _____
3. 5.678; 725; 0.974; 67.26; 7.87; 0.387   _____

Circle the numbers in each row that have a 2 in the *thousandths* place. Add those numbers with your calculator and write your answer in the blank.

1. 2,849.1; 0.234; 346.432; 0.482; 2.372   _____
2. 0.372; 2,740.452; 2,748; 12.67; 48.762   _____
3. 12.732; 0.0832; 4.321; 4.312; 0.782   _____

Circle the numbers in each row that have a 4 in the *ones* place. Add those numbers with your calculator and write your answer in the blank.

1. 894.67; 49.87; 14.16; 1,784.6; 347.44   _____
2. 426.43; 4,009.24; 344.8; 24.87; 62.4   _____
3. 98.4; 104.8; 543.9; 4,628.4; 364.14   _____

ROUNDING NUMBERS

It is sometimes necessary to simplify, or *round*, numbers to make them easier to use.

For example, when some states figure sales tax, the tax may be $6\frac{1}{2}$¢ on each $1 purchase. There is no $\frac{1}{2}$¢ coin; so if you purchase an article for $1.00, and a $6\frac{1}{2}$¢ tax is added, the total is rounded to $1.07. Rounding is often used in estimating numbers.

Rounding Whole Numbers

Example 1   376 rounded to the nearest ten would be 380.

If the number in the ones place is 5 or more, add 1 to the number in the tens place and put a 0 in the ones place. Similarly, 376 rounded to the nearest hundred is 400.

Example 2   374 rounded to the nearest ten would be 370.

If the number in the ones place is less than 5, the number in the tens place will remain the same. Put a 0 in the ones place. Similarly, 234 rounded to the nearest hundred is 200.

Example 3   8,620 rounded to the nearest thousand would be 9,000.

Round the following numbers to the nearest value indicated.

Ten

1. 48 ___50___
2. 159 _____
3. 3,273 _____
4. 64 _____
5. 85 _____

Hundred

6. 485 ___500___
7. 34,973 _____
8. 2,137 _____
9. 623 _____
10. 860 _____

Thousand

11. 2,371 ___2,000___
12. 12,862 _____
13. 196,584 _____
14. 10,699 _____
15. 6,486 _____

Rounding Decimals

When rounding decimals, use the same process as when rounding whole numbers. Remember to keep the decimal point in the correct place.

Example 1   0.284 rounded to the nearest tenth is 0.3.

Example 2   0.284 rounded to the nearest hundredth is 0.28.

Example 3   0.285 rounded to the nearest hundredth is 0.29.

Example 4   0.2856 rounded to the nearest thousandth is 0.286.

Round the following numbers to the nearest value indicated.

Tenth	Hundredth	Thousandth
1. 2.26 __2.3__	11. 0.862 __0.86__	21. 0.0397 __0.040__
2. 0.48 _____	12. 2.4956 _____	22. 4.2843 _____
3. 0.7921 _____	13. 0.059 _____	23. 0.9654 _____
4. 0.06 _____	14. 8.642 _____	24. 8.6681 _____
5. 0.333 _____	15. 0.229 _____	25. 0.4567 _____
6. 2.16 _____	16. 0.751 _____	26. 8.0001 _____
7. 0.98 _____	17. 8.088 _____	27. 8.0005 _____
8. 0.44 _____	18. 0.555 _____	28. 0.1772 _____
9. 0.83 _____	19. 0.906 _____	29. 0.5055 _____
10. 0.67 _____	20. 1.009 _____	30. 4.2222 _____

Rounding Money

The method for rounding money is no different than for rounding any other number.

Example 1 $348.39 to the nearest *dollar* would be $348.

The number 8 is in the one dollar place, and 0.39 is less than 0.50. Therefore, the 0.39 is dropped and the $348 remains the same.

Example 2 $357 to the nearest *ten dollars* would be $360.

Round the following to the nearest amount indicated.

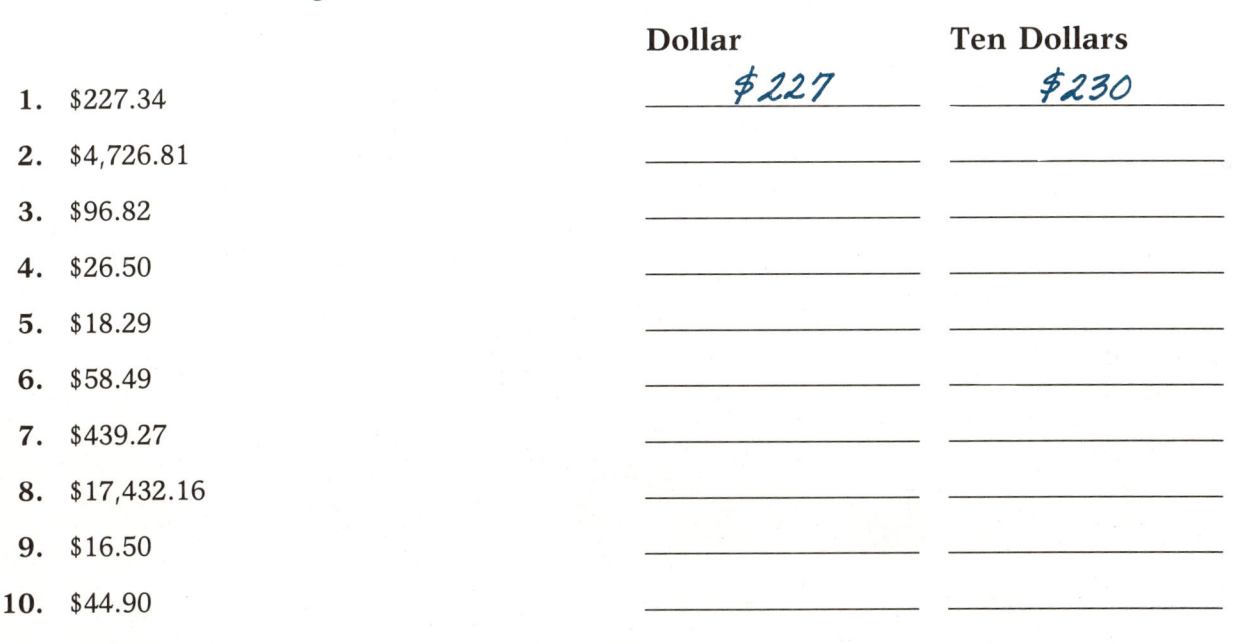

	Dollar	Ten Dollars
1. $227.34	$227	$230
2. $4,726.81	_____	_____
3. $96.82	_____	_____
4. $26.50	_____	_____
5. $18.29	_____	_____
6. $58.49	_____	_____
7. $439.27	_____	_____
8. $17,432.16	_____	_____
9. $16.50	_____	_____
10. $44.90	_____	_____

UNIT 1 ■ THE CALCULATOR AND A REVIEW OF THE BASICS

Example 3 $57.888 to the nearest penny (hundredth) would be $57.89.

Round the following to the nearest hundredth.

1. $0.671 _____
2. $2.225 _____
3. $3.564 _____
4. $0.467 _____
5. $5.396 _____
6. $21.999 _____

Rounding with the Calculator

An understanding of rounding is essential when you use a calculator. Have you ever noticed something like this when using a calculator?

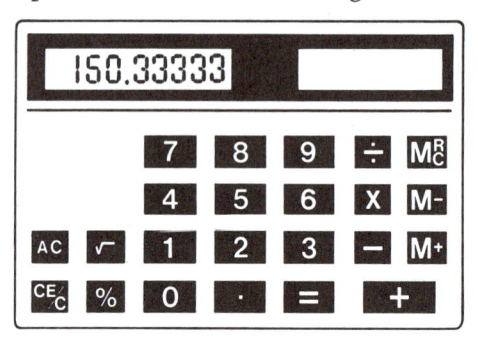

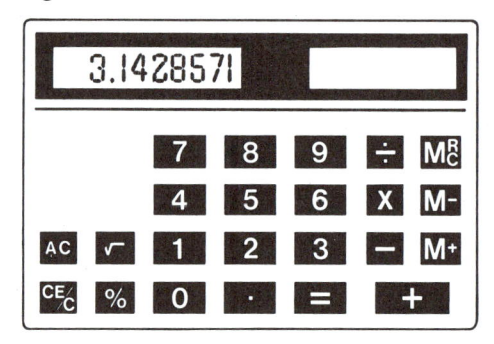

150.33333   or   3.1428571

Enter the following on your calculator and write what you see.

Do This   **See This**

1. 16 ÷ 3 = _____
2. 132.43 × 0.07 = _____
3. 17,247 ÷ 13 = _____
4. 347.63 × 0.049 = _____

Answer the following questions and round the calculator answer as indicated.

1. If oil filters are sold 3 for $10, how much does 1 oil filter cost? (Round to the nearest hundredth.) _____

2. A portable stereo costs $125.78. How much is a 6% sales tax? (6% = 0.06. Multiply $125.78 by 0.06 and round your answer to the nearest hundredth.)* _____

3. Twenty-one students distributed 5,860 fliers announcing a flea market. About how many fliers did each student distribute? (Round to the nearest whole number.) _____

4. Four new tires cost $285.59. How much is a $5\frac{1}{2}$% sales tax? (Multiply $285.59 by 0.055 and round your answer to the nearest hundredth.)* _____

*Changing percents to decimals will be discussed on page 37.

DECIMALS

Adding, subtracting, multiplying, and dividing decimals is the same as adding, subtracting, multiplying and dividing whole numbers. You just have to remember to keep track of the decimal point.

Adding and Subtracting Decimals

When adding or subtracting decimals without a calculator, you should arrange the numbers in a column with the decimal points lined up. When a number is written without a decimal point, the decimal point is at the right of the last digit of the number. Rewrite the number with the decimal point at the right and add as many zeros as needed so the columns of digits line up. Then add or subtract the numbers. The decimal point in the answer lines up with the decimal points above it.

Example 1

$15.75 + $8.95 + $5 =

$15.75
  8.95
+ 5.00

$29.70

Example 2

$28 − $17.72 =

$28.00
− 17.72

$10.28

Solve the following problems without using your calculator.

1. $12 + $0.89 + 6.93 = _____

2. $500 − $235.29 = _____

3. $4,592 − $529.63 = _____

UNIT 1 ■ THE CALCULATOR AND A REVIEW OF THE BASICS

Solve the following problems using your calculator.

1. Sonya buys a dress that costs $28.98. How much change does she get from 3 ten-dollar bills?

2. Andrew buys groceries that cost $1.79, $0.59, $1, $0.99, and $2.56. What is the total cost of the groceries?

3. Linda and Cecil went to a concert. The tickets were $15.50 each, and they paid with a $50 bill. How much change did they receive?

4. Jeremy has chosen items with the following prices: $10, $7.95, $0.39, and $1.55. What is the total cost of the items?

Multiplying Decimals

When multiplying decimals without using a calculator, multiply as though there are no decimal points. Then count the total number of places to the right of the decimal points in the problem. Your answer must have the same number of places to the right of the decimal point. Always count from right to left when placing the decimal point in the answer.

Example 1

456 × 5.7 =

```
     4 5 6
  ×    5.7    one place
   ------
   3 1 9 2
   2 2 8 0
   ------
   2,5 9 9.2  one place
```

$$0.50 \times 0.7 = 0.350$$

Example 2

5.28 × 3.5 =

```
     5.2 8    two places
  ×    3.5    one place
   ------
   2 6 4 0
   1 5 8 4
   ------
   1 8.4 8 0  three places
```

14   UNIT 1 ■ THE CALCULATOR AND A REVIEW OF THE BASICS

If you don't have enough decimal places in the answer, write as many zeros as needed in front of the answer.

Example 3

0.0056 × 0.2 =

```
  0.0 0 5 6    four places
×       0.2    one place
  0.0 0 1 1 2  five places
```

Multiply the following numbers without using your calculator.

1. 32.5 × 32 =

2. 3.7 × 28 =

3. 3.98
   × 7.6

4. 3.2 × 0.05 =

5. 0.0087
   × 0.09

6. 0.065 × 0.05 =

Solve the following problems using your calculator.

1. Wayne saves $18.50 a week. How much can he save in a year? (There are 52 weeks in 1 year.) _____

2. Sue's car insurance is $58.67 a month. She wants to pay once every 6 months. How much will she owe for 6 months? _____

UNIT 1 ■ THE CALCULATOR AND A REVIEW OF THE BASICS

Dividing Decimals

When dividing a decimal number by a whole number, write the decimal point for the answer directly above the decimal point in the dividend. Then divide. Be sure to keep your columns lined up so you don't lose your place.

Example 1

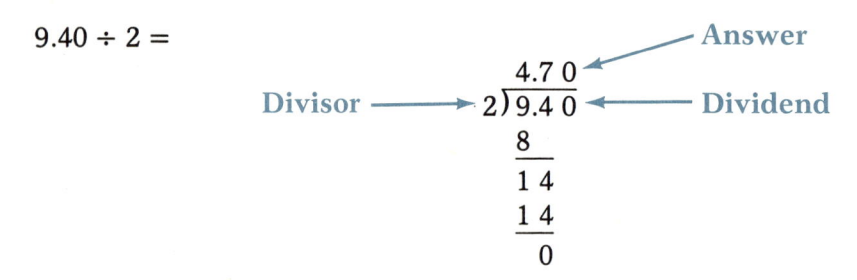

$9.40 \div 2 =$

```
           4.70  ← Answer
Divisor → 2)9.40  ← Dividend
           8
           ―
           1 4
           1 4
           ―
             0
```

Divide the following numbers without using a calculator.

1. $2\overline{)7.54}$

2. $64.2 \div 3 =$

When dividing a decimal number by another decimal number, first make the divisor a whole number by moving the decimal point to the right. Move the decimal point in the dividend the same number of places to the right. Write the decimal point for the answer directly above the decimal point in the dividend. Then divide.

Example 2

$940 \div 0.2 =$

```
              4 7 0.
    0.2.)9 4 0.
         8
         ―
         1 4
         1 4
         ―
           0
```

$0.5.\overline{)4.5.}$

16   UNIT 1 ■ THE CALCULATOR AND A REVIEW OF THE BASICS

Example 3

9.045 ÷ 0.15 =

```
          6 0.3
0.15.)9.0 4.5
       9 0
          4
          0
          4 5
          4 5
              0
```

Divide the following numbers without using a calculator.

1. 0.2)6.4 4

2. 67.68 ÷ 9.4 =

3. 0.42)8.4 8 4

4. 0.3024 ÷ 0.84 =

UNIT 1 ■ THE CALCULATOR AND A REVIEW OF THE BASICS

When dividing a whole number by a decimal, you must *write* a decimal point and the necessary number of zeros in the dividend.

Example 4

$$940 \div 0.2 =$$

```
           4,7 0 0
    0.2.)9,4 0.0.
           8
          ---
           1 4
           1 4
          ---
             0
```

Example 5

$$9,045 \div 0.15 =$$

```
              6 0,3 0 0
     0.15.)9,0 4 5.0 0.
              9 0
             ----
                 4
                 4
               ---
                 4 5
                 4 5
                ---
                   0
```

Divide the following numbers without using a calculator.

1. $0.2 \overline{)6,472}$

2. $2,408 \div 4.3 =$

3. $0.20 \overline{)6,040}$

4. $6,816 \div 0.48 =$

Solve the following problems using your calculator.

1. Dorothy used 24.5 gallons of fuel to drive 490 miles. How many miles per gallon did she get? _____

2. Teresa makes $3.35 an hour. How many hours will it take her to earn $207.70? _____

3. Ryan wants to buy a car that costs $762.50. If he can save $30.50 a week, how many weeks will it take him to save enough money to buy the car? _____

4. How many pounds of apples can you buy for $4.72, if they cost 59¢ a pound? _____

Understanding Multiplying and Dividing by Less Than One

If you understand the kind of answer you can expect to get, multiplying and dividing decimals makes sense.

When you *multiply* 100 by 0.25, the answer is *smaller* than 100.

Example 1

100 × 0.25 =

```
    1 0 0
  × 0.2 5
  -------
    5 0 0
  2 0 0
  -------
  2 5.0 0
```

Solve these paired problems and compare the answers.

1. Multiply $20 by 4 _____   Multiply $20 by 0.40 _____

2. Multiply $15 by 3 _____   Multiply $15 by 0.30 _____

3. Multiply 11 by 0.5 _____   Multiply 0.11 by 0.5 _____

When you *divide* 100 by 0.25, the answer is *larger* than 100.

Example 2

100 ÷ 0.25 =

```
          4 0 0
 0.25)1 0 0.0 0
        1 0 0
        -----
            0
```

UNIT 1 ■ THE CALCULATOR AND A REVIEW OF THE BASICS

Solve the following paired problems and compare the answers.

1. Divide $20 by $4 _____ Divide $20 by $0.40 _____

2. Divide $15 by $3 _____ Divide $15 by $0.30 _____

3. Divide 11 by 0.5 _____ Divide 0.11 by 0.5 _____

FRACTIONS

When working with fractions, you will need to know the meanings of the following terms.

Numerator. Top number of fraction: $\frac{①}{2}$

Denominator. Bottom number of fraction: $\frac{2}{③}$

Proper fraction. Numerator is smaller than denominator: $\frac{5}{6}$

Improper fraction. Numerator is equal to or larger than denominator: $\frac{3}{3}$ or $\frac{5}{3}$

Mixed number. Whole number plus a fraction: $2\frac{4}{7}$

Comparing Size of Fractions

If the numerators of fractions are the same, the larger denominator shows the *smaller* value.

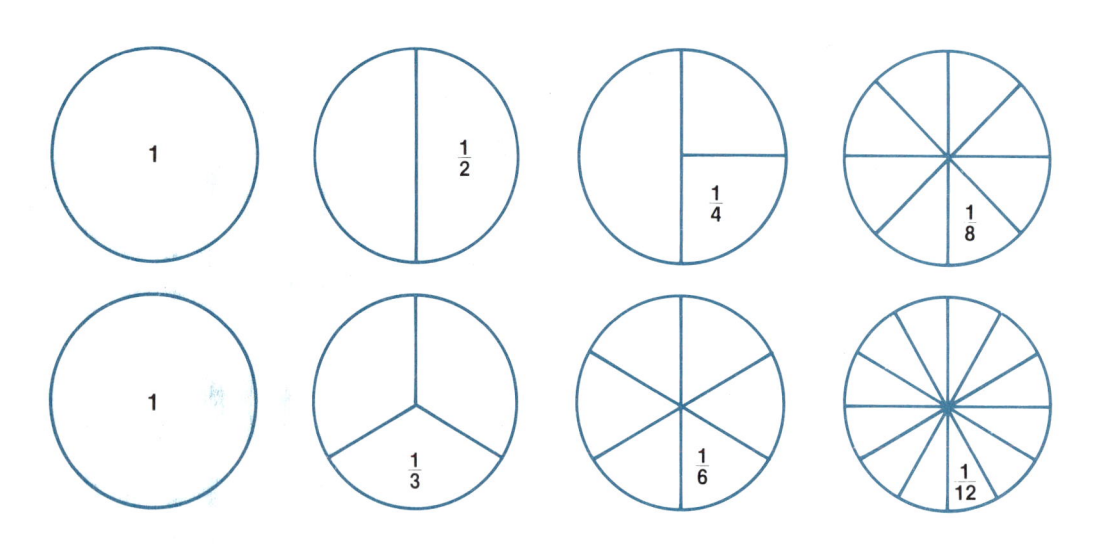

Put the following fractions in order beginning with the smallest and working up to the largest in each row below.

Row 1: $\frac{1}{3}, \frac{1}{12}, \frac{1}{6}, 1, \frac{2}{3}$ _____

Row 2: $\frac{1}{3}, \frac{3}{4}, \frac{1}{8}, \frac{1}{2}, \frac{1}{6}, \frac{11}{12}, \frac{5}{6}$ _____

Reducing Fractions

Reducing a fraction means making the numerator and denominator smaller. The reduced fraction is equal to the original fraction. To reduce a fraction find a number that will divide evenly into both the numerator and the denominator.

Example

$\frac{6}{8}$ reduces to $\frac{3}{4}$ $\frac{6 \div 2}{8 \div 2} = \frac{3}{4}$

$\frac{6}{8}$ of a circle is the same as $\frac{3}{4}$ of a circle.

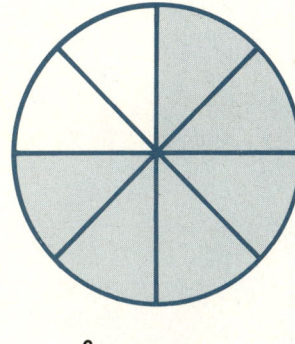

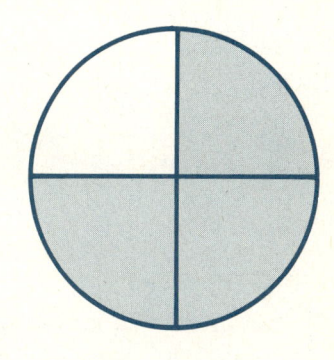

$\frac{6}{8}$ of a circle $\frac{3}{4}$ of a circle

HINT: When 1 is the only number that will divide evenly into the numerator and denominator, the fraction cannot be reduced.

When a fraction can no longer be reduced, the fraction is said to be *reduced to lowest terms*. Fractional answers to problems should usually be reduced to lowest terms.

Reduce the following fractions to lowest terms.

1. $\frac{2}{4} =$ _____

2. $\frac{12}{16} =$ _____

3. $\frac{200}{500} =$ _____

4. $\frac{3}{9} =$ _____

5. $\frac{10}{25} =$ _____

6. $\frac{9}{36} =$ _____

7. $\frac{4}{16} =$ _____

8. $\frac{6}{9} =$ _____

9. $\frac{5}{15} =$ _____

10. $\frac{8}{12} =$ _____

11. $\frac{100}{250} =$ _____

12. $\frac{250}{500} =$ _____

UNIT 1 ■ THE CALCULATOR AND A REVIEW OF THE BASICS

Changing Improper Fractions to Mixed Numbers

To change an improper fraction to a mixed number, divide the denominator into the numerator. Put the remainder over the divisor.

Example

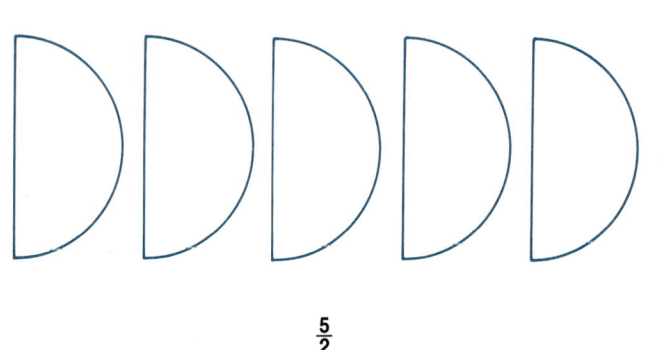

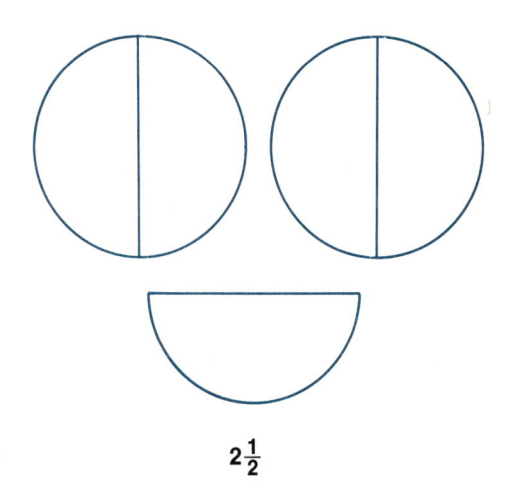

$$\frac{5}{2} = 2\overline{)5} = 2\frac{1}{2}$$

Five half-circles equal two and a half circles.

Change the following improper fractions to mixed numbers, and reduce to lowest terms when possible.

1. $\frac{27}{10} =$ _____
2. $\frac{9}{4} =$ _____
3. $\frac{500}{200} =$ _____
4. $\frac{10}{3} =$ _____
5. $\frac{13}{6} =$ _____

6. $\frac{11}{2} =$ _____
7. $\frac{7}{2} =$ _____
8. $\frac{4}{2} =$ _____
9. $\frac{12}{12} =$ _____
10. $\frac{400}{200} =$ _____

11. $\frac{4}{4} =$ _____
12. $\frac{9}{8} =$ _____
13. $\frac{20}{5} =$ _____
14. $\frac{250}{100} =$ _____
15. $\frac{16}{6} =$ _____

Changing Mixed Numbers to Improper Fractions

To change a mixed number to an improper fraction, follow these steps: Multiply the denominator of the fraction by the whole number, then add the numerator of the fraction. Write this sum over the original denominator.

Example Two and one-third equals seven-thirds.

$$2\frac{1}{3} = \frac{7}{3}$$

$2 \times 3 + 1 = 7$, which is the numerator of the improper fraction. Place the numerator over the original denominator.

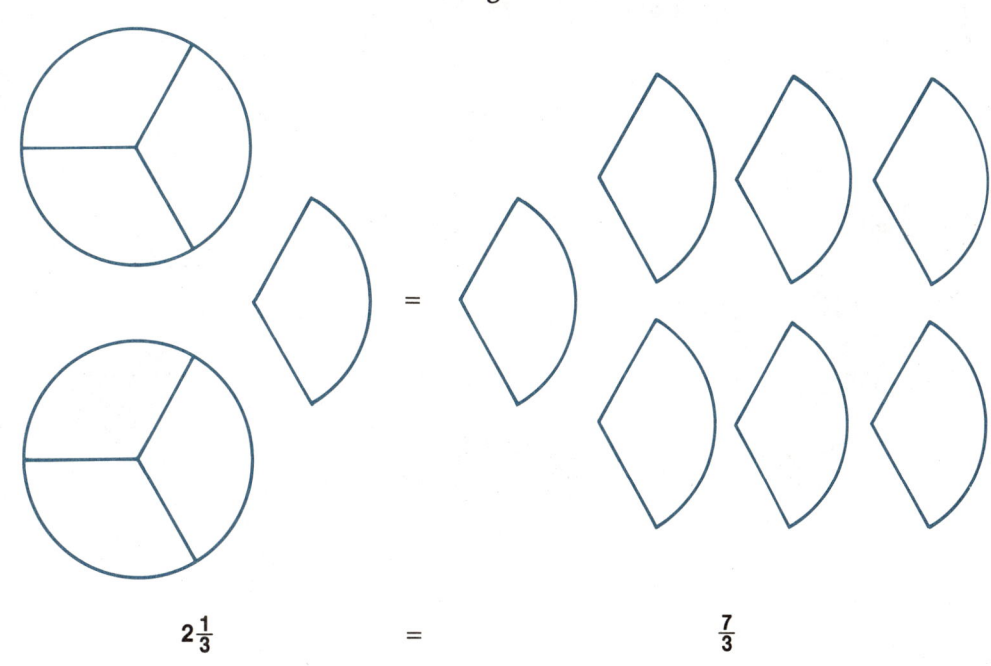

$$2\frac{1}{3} = \frac{7}{3}$$

HINT: The denominator of the improper fraction remains the same.

Change the following mixed numbers to improper fractions.

1. $5\frac{3}{4} =$ _____

2. $2\frac{7}{8} =$ _____

3. $3\frac{1}{2} =$ _____

4. $4\frac{1}{3} =$ _____

5. $1\frac{3}{5} =$ _____

6. $7\frac{5}{6} =$ _____

7. $5\frac{1}{8} =$ _____

8. $7\frac{1}{2} =$ _____

9. $2\frac{2}{7} =$ _____

10. $3\frac{3}{10} =$ _____

11. $2\frac{2}{3} =$ _____

12. $3\frac{1}{5} =$ _____

UNIT 1 ■ THE CALCULATOR AND A REVIEW OF THE BASICS

Multiplying Fractions

When multiplying fractions, multiply numerators and denominators as shown in the examples. Remember to reduce your answer to lowest terms.

Example 1

$$\frac{2}{3} \times \frac{3}{4} = \frac{2 \times 3}{3 \times 4} = \frac{6}{12} = \frac{1}{2}$$

Example 2

$$5 \times \frac{3}{7} = \frac{5}{1} \times \frac{3}{7} = \frac{5 \times 3}{1 \times 7} = \frac{15}{7} = 2\frac{1}{7}$$

HINT: Write a whole number as a fraction by placing the whole number over 1.

Multiply the following fractions.

1. $\frac{1}{4} \times \frac{2}{3} =$ _____

2. $\frac{3}{4} \times \frac{1}{2} =$ _____

3. $8 \times \frac{1}{10} =$ _____

4. $\frac{5}{6} \times \frac{3}{4} =$ _____

When multiplying mixed numbers, first change them to improper fractions. Then multiply the numerators and denominators.

Example 3

$2\frac{1}{3} \times 1\frac{1}{2}$

Step 1

$2\frac{1}{3} = \frac{7}{3}$ $(3 \times 2 + 1 = 7)$

$1\frac{1}{2} = \frac{3}{2}$ $(2 \times 1 + 1 = 3)$

Step 2

$\frac{7}{3} \times \frac{3}{2} = \frac{21}{6} = 3\frac{3}{6} = 3\frac{1}{2}$

Canceling

The following is a *shortcut for multiplying fractions*.

Step 1 Look at the *opposite* numerators and denominators. Can they be divided evenly by the same number?

Step 2 Divide both the numerator and denominator by the same number.

Example

$\frac{6}{7} \times \frac{1}{9} = \frac{\cancel{6}^{2}}{7} \times \frac{1}{\cancel{9}_{3}} = \frac{2}{21}$ Both 6 and 9 are divisible by 3.

Solve the following problems. You decide whether you wish to take the canceling shortcut.

1. $2\frac{1}{2} \times 6\frac{2}{5} = \frac{\cancel{5}^1}{\cancel{2}_1} \times \frac{\cancel{32}^{16}}{\cancel{5}_1} = \frac{16}{1} = 16$ __16__

2. $5\frac{1}{3} \times 3 =$ _____

3. $\frac{2}{5} \times \frac{1}{4} =$ _____

4. $2\frac{1}{5} \times 6\frac{1}{4} =$ _____

5. $5 \times 9\frac{3}{10} =$ _____

6. $2\frac{2}{3} \times \frac{1}{6} =$ _____

7. $5\frac{1}{3} \times 3 =$ _____

8. $\frac{1}{8} \times 1\frac{1}{7} =$ _____

UNIT 1 ■ THE CALCULATOR AND A REVIEW OF THE BASICS

9. If you plan to sew 8 shirts, and each shirt requires $3\frac{1}{4}$ yards of of fabric, how many yards of fabric must you buy? _____

10. How many feet of wood are needed to make 21 shelves if each shelf measures $6\frac{2}{3}$ feet? _____

11. Find the price of $1\frac{1}{2}$ dozen eggs at 68¢ per dozen. _____

12. A house worth $79,500 is taxed at $\frac{2}{3}$ of its value. On what dollar amount will the owners pay taxes? _____

Dividing Fractions

When dividing fractions and mixed numbers, turn the second fraction upside down and multiply the fractions.

Example 1

$\frac{1}{4} \div \frac{2}{3} = \frac{1}{4} \times \frac{3}{2} = \frac{3}{8}$

Example 2

$2\frac{1}{3} \div \frac{1}{4} = \frac{7}{3} \times \frac{4}{1} = \frac{28}{3} = 9\frac{1}{3}$

Example 3

$$10 \div \frac{1}{5} = \frac{10}{1} \times \frac{5}{1} = \frac{50}{1} = 50$$

> **HINT:** This really makes sense if you think about it. How many $\frac{1}{5}$s are in the whole number 10? 50 of them!

Example 4

$$\frac{1}{5} \div 10 = \frac{1}{5} \div \frac{10}{1} = \frac{1}{5} \times \frac{1}{10} = \frac{1}{50}$$

> **HINT:** If the second number is a whole number, write the second number as a fraction by putting it over 1. Then turn the fraction upside down.

Solve the following problems.

1. $4 \div \frac{4}{5} =$ _____

2. $2\frac{1}{3} \div 1\frac{2}{3} =$ _____

3. $2\frac{1}{4} \div \frac{1}{2} =$ _____

4. $2\frac{1}{2} \div 2 =$ _____

5. If a plane travels 850 miles in $2\frac{1}{2}$ hours, how many miles per hour does it travel? _____

6. If you pay 63¢ for $1\frac{3}{4}$ pounds of apples, how much does 1 pound cost? _____

7. How many pieces of wood $\frac{1}{3}$ foot long can be cut from a board $6\frac{2}{3}$ feet long? _____

Understanding Multiplying and Dividing by Less Than One

Just as in multiplying and dividing decimals, it's important to understand the kind of answer you get when you multiply and divide fractions.

When multiplying 2 by $\frac{1}{4}$, you will find that the answer is *smaller* than 2.

Example 1

$$2 \times \frac{1}{4} = \frac{2}{1} \times \frac{1}{4} = \frac{2}{4} = \frac{1}{2}$$

Solve the following paired problems and compare the answers.

1. Multiply 10 by 8 = _____   Multiply 10 by $\frac{1}{8}$ = _____

2. Multiply 15 by 3 = _____   Multiply 15 by $\frac{1}{3}$ = _____

3. Multiply 12 by 4 = _____   Multiply 12 by $\frac{1}{4}$ = _____

When dividing 24 by $\frac{1}{4}$, you will find that the answer is *larger* than 24.

Example 2

$$24 \div \frac{1}{4} = \frac{24}{1} \times \frac{4}{1} = 96$$

Solve these paired problems and compare the answers.

1. Divide 500 by 10 = _____ Divide 500 by $\frac{1}{10}$ = _____

2. Divide 24 by 3 = _____ Divide 24 by $\frac{1}{3}$ = _____

3. Divide 36 by 9 = _____ Divide 36 by $\frac{1}{9}$ = _____

Adding and Subtracting Fractions

To add or subtract fractions with the *same* denominators, add or subtract the numerators. Write the answer with the common denominator.

Example 1

$$\begin{array}{r} \frac{1}{10} \\ + \frac{3}{10} \\ \hline \frac{4}{10} \end{array}$$

Example 2

$$\begin{array}{r} \frac{2}{3} \\ - \frac{1}{3} \\ \hline \frac{1}{3} \end{array}$$

To add or subtract fractions with *different* denominators, you must first find a common denominator. To find a common denominator, find a number that both denominators divide into evenly. After finding the common denominator, you must make equal, or *equivalent*, fractions before adding or subtracting the fractions.

Equivalent fractions are the same size.

$\frac{1}{2} = \frac{2}{4}$ $\frac{1}{2} = \frac{1 \times 2}{2 \times 2} = \frac{2}{4}$

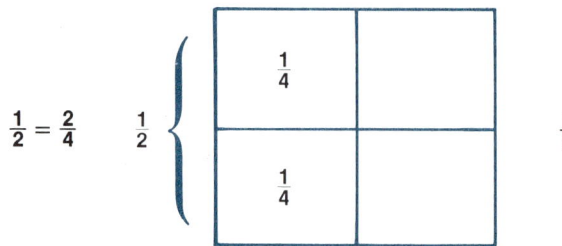

$\frac{3}{4} = \frac{6}{8}$ $\frac{3}{4} = \frac{3 \times 2}{4 \times 2} = \frac{6}{8}$

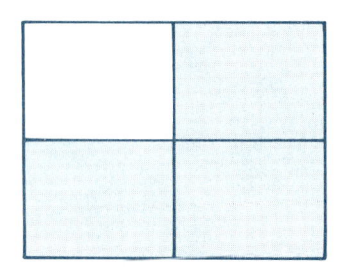

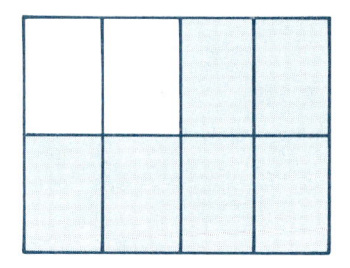

HINT: Multiply the numerator and denominator by the same number.

Example 3

$\frac{1}{2} = \frac{5}{10}$ Both 2 and 5 divide
$+\frac{2}{5} = \frac{4}{10}$ evenly into 10.

Example 4

$\frac{3}{4} = \frac{6}{8}$ Both 4 and 8 divide
$-\frac{2}{8} = \frac{2}{8}$ evenly into 8.

HINT: It's easier to add or subtract if you find the *lowest* common denominator. This is the *smallest* number that both denominators will divide into evenly.

Find a common denominator for the following pairs of fractions and make equivalent fractions.

1. $\frac{2}{3} = $ _____
   $\frac{1}{6} = $ _____

2. $\frac{1}{2} = $ _____
   $\frac{2}{3} = $ _____

3. $\frac{2}{5} = $ _____
   $\frac{1}{3} = $ _____

4. $\frac{2}{3} = $ _____
   $\frac{1}{4} = $ _____

5. $\frac{1}{2} = $ _____
   $\frac{1}{5} = $ _____

Study the following examples of adding and subtracting fractions with unlike denominators.

Example 5

$$\frac{1}{2} = \frac{5}{10}$$
$$+\frac{2}{5} = \frac{4}{10}$$
$$\overline{\frac{9}{10}}$$

Example 6

$$\frac{3}{4} = \frac{6}{8}$$
$$-\frac{2}{8} = \frac{2}{8}$$
$$\overline{\frac{4}{8} = \frac{1}{2}}$$

Example 7

$$6\frac{5}{6} = 6\frac{10}{12}$$
$$+2\frac{3}{4} = 2\frac{9}{12}$$
$$\overline{8\frac{19}{12} = 8 + \frac{19}{12} = 8 + 1\frac{7}{12} = 9\frac{7}{12}}$$

> **HINT:** If the fraction part of a mixed number is an improper fraction, change the improper fraction to a mixed number and add it to the whole number.

Add and subtract the following fractions and mixed numbers. Be sure to find a common denominator when necessary and reduce your answer when possible.

1. $\frac{7}{16}$
   $-\frac{2}{8}$

2. $\frac{5}{6}$
   $+\frac{3}{4}$

3. $4\frac{2}{6}$
   $-\frac{1}{3}$

4. $1\frac{3}{12}$
   $+3\frac{1}{6}$

5. $2\frac{2}{3}$
   $-1\frac{2}{5}$

6. $\frac{3}{5}$
   $4\frac{1}{10}$
   $+2\frac{2}{5}$

UNIT 1 ■ THE CALCULATOR AND A REVIEW OF THE BASICS

7. You have a piece of material $9\frac{1}{2}$ feet long. If you cut a piece $3\frac{1}{4}$ feet long, how much material will you have left? _____

8. If you worked $1\frac{1}{4}$ hours on Monday, $\frac{1}{2}$ hour on Wednesday, and $2\frac{3}{4}$ hours on Friday, how many hours did you work altogether? _____

Borrowing When Subtracting Fractions

Sometimes you must *borrow* when you subtract fractions and mixed numbers. For example, you can't subtract $1\frac{2}{3}$ from $3\frac{1}{3}$ because $\frac{2}{3}$ is larger than $\frac{1}{3}$. Follow the steps in the example to see how to borrow when subtracting fractions and mixed numbers.

Example 1

$$3\frac{1}{3}$$
$$-1\frac{2}{3}$$

Step 1 Borrow 1 from the whole number. Change the borrowed 1 to its equivalent fraction, $\frac{3}{3}$, and add $\frac{3}{3}$ to $\frac{1}{3}$.

$$3\frac{1}{3} = 2\frac{1}{3} + 1 = 2\frac{1}{3} + \frac{3}{3} = 2\frac{4}{3}$$

Step 2 Subtract $1\frac{2}{3}$ from $2\frac{4}{3}$.

$$2\frac{4}{3}$$
$$-1\frac{2}{3}$$
$$\overline{1\frac{2}{3}}$$

HINT: 1 can be written as a fraction that has the same numerator and denominator; $\frac{2}{2}$, $\frac{3}{3}$, and $\frac{11}{11}$ are all equivalent to 1.

Example 2

$$4\frac{1}{3} = 4\frac{4}{12} = 3\frac{16}{12}$$
$$-2\frac{3}{4} = 2\frac{9}{12} = 2\frac{9}{12}$$
$$\overline{\phantom{-2\frac{3}{4} = 2\frac{9}{12} =\ } 1\frac{7}{12}}$$

HINT: Find a common denominator before borrowing.

Solve the following problems.

1. $5\frac{1}{3}$
   $-2\frac{2}{3}$

2. $5\frac{3}{4}$
   $-2\frac{5}{6}$

3. 2
   $-\frac{2}{3}$

4. A $4\frac{1}{4}$ pound chicken weighed $3\frac{3}{8}$ pounds when dressed. What was the loss in weight? _____

5. Two months ago Sandra weighed $130\frac{1}{4}$ pounds. Now she weighs $123\frac{3}{4}$ pounds. How many pounds did she lose? _____

UNIT 1 ■ THE CALCULATOR AND A REVIEW OF THE BASICS

Add, subtract, multiply, or divide to solve the following problems.

1. Max needs $2\frac{1}{4}$ cups of flour for cookies and $\frac{1}{2}$ cup of flour for biscuits. How much flour does he need altogether? _____

2. Dennis has 640 marbles. He is giving $\frac{3}{8}$ of them to his brother. How many marbles will he give to his brother? _____

3. Emily is delivering cookies. She must leave $\frac{2}{3}$ of the 72 cookies at Tim's house. How many cookies will she leave? _____

4. If a pilot flies 665 miles in $1\frac{3}{4}$ hours, what is the pilot's speed per hour? _____

5. You plan to make curtains for a friend's van. You need $1\frac{1}{2}$ yards for the back, $2\frac{1}{4}$ yards for one side, $3\frac{3}{8}$ yards for the other side, and $1\frac{3}{4}$ yards for the front. How many yards do you need altogether? _____

6. If you need $2\frac{1}{2}$ pounds of hamburger for one recipe and $4\frac{1}{3}$ pounds for another recipe, how many pounds do you need altogether? _____

7. If a person drives 455 miles and it takes her $8\frac{3}{4}$ hours, how many miles per hour did she drive? _____

8. Rene is sewing a tablecloth and matching napkins. If she has $10\frac{1}{2}$ yards of fabric and uses $6\frac{3}{4}$ yards for the tablecloth, how much fabric is left for the napkins? _____

9. A $15\frac{1}{2}$-pound live turkey weighed $12\frac{1}{4}$ pounds when ready to cook. What was the weight loss? _____

10. How much will a 150-mile trip cost at $12\frac{1}{2}$¢ per mile? _____

UNIT 1 ■ THE CALCULATOR AND A REVIEW OF THE BASICS

EQUIVALENTS: FRACTIONS = DECIMALS = PERCENTS

A value less than one may be written as a fraction, a decimal, or a percent. *Percent* means something has been divided into 100 parts. For example, 25% means 25 of 100 or $\frac{25}{100}$, or 0.25.

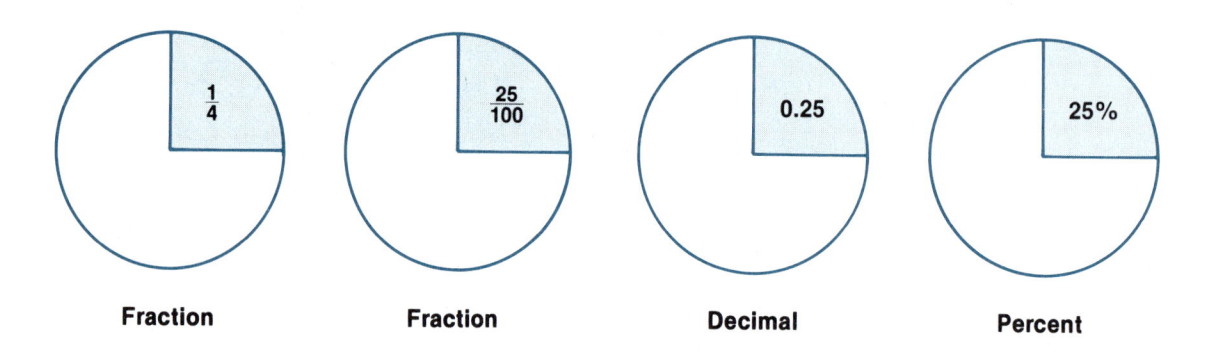

$\frac{1}{4}$	$\frac{25}{100}$	0.25	25%
Fraction	Fraction	Decimal	Percent

In the circles shown above, the shaded parts are equal.

$$\frac{1}{4} = \frac{25}{100} = 0.25 = 25\%$$

When we talk about money, we say either half a dollar or 50¢. This is the same value as $\frac{50}{100}$ or 50% of a dollar.

$$\frac{1}{2} = \frac{50}{100} = 0.50 = 50\%$$

Changing from Decimals to Percents

To change a decimal to a percent, move the decimal point *two* places to the right. Write the percent sign to the right of the last digit.

Examples

0.65 = 65%
0.20 = 20%
0.025 = 2.5%
2.1 = 210%

Change these decimals to percents.

1. 0.10 = _____
2. 0.056 = _____
3. 5.8 = _____
4. 0.07 = _____

5. 0.79 = _____
6. 0.06 = _____
7. 0.25 = _____
8. 1 = _____

Changing from Percents to Decimals

To change a percent to a decimal, move the decimal point *two* places to the left. Drop the percent sign.

Examples

15% = 0.15
25% = 0.25
10% = 0.10
6% = 0.06
5.5% = 0.055

> **HINT:** Write a zero between the decimal point and the number when there is a one-digit percent.
>
> **7% = 0.07**

Change the following percents to decimals.

1. 5% = _____
2. 10% = _____
3. 250% = _____
4. 3.5% = _____
5. 39% = _____
6. 3% = _____
7. 100% = _____
8. 85% = _____

Changing from Percents to Fractions

To change a percent to a fraction, place the percent over 100 and remove the percent sign (%). Reduce the fraction when possible.

Examples

$25\% = \frac{25}{100} = \frac{1}{4}$

$5\% = \frac{5}{100} = \frac{1}{20}$

> **HINT:** Remember that when you reduce a fraction, you must find a number that will divide evenly into both the numerator and the denominator.
>
> $\frac{25}{100} = \frac{25 \div 5}{100 \div 5} = \frac{1}{4}$

Change from percents to fractions and reduce when possible.

1. 50% = _____
2. 10% = _____
3. 3% = _____
4. 15% = _____
5. 30% = _____
6. 8% = _____
7. 80% = _____
8. 77% = _____

UNIT 1 ■ THE CALCULATOR AND A REVIEW OF THE BASICS

Changing from Fractions to Percents

To change a fraction to a percent, first divide the denominator into the numerator. Change the decimal that results to a percent.

Example Change $\frac{1}{4}$ to a percent.

Step 1

Divide the denominator into the numerator.

$$\begin{array}{r} 0.25 \\ 4\overline{)1.00} \\ \underline{8} \\ 20 \\ \underline{20} \\ 0 \end{array}$$

Step 2

Move the decimal point two places to the *right* in your answer. Write a percent sign.

$0.25 = 25\%$

Change from fractions to percents.

1. $\frac{1}{10} =$ _____
2. $\frac{1}{2} =$ _____
3. $\frac{3}{2} =$ _____

4. $\frac{3}{4} =$ _____
5. $\frac{2}{5} =$ _____
6. $\frac{7}{10} =$ _____

Using your calculator, change from fractions to decimals. Then change the decimals to percents.

	Fraction	Decimal	Percent
1.	$\frac{3}{8}$	0.375	37.5%
2.	$\frac{9}{20}$		
3.	$\frac{3}{5}$		
4.	$\frac{1}{8}$		
5.	$\frac{7}{20}$		
6.	$\frac{5}{8}$		
7.	$\frac{1}{5}$		
8.	$\frac{3}{10}$		

Changing from Decimals to Fractions

To change a decimal to a fraction, read the decimal number. Then write the number as a fraction. (Refer to the chart on page 4.)

Example 1

0.7 is read seven tenths.

$0.7 = \frac{7}{10}$

Example 2

0.35 is read thirty-five hundredths.

$0.35 = \frac{35}{100}$

Example 3

0.002 is read two thousandths.

$0.002 = \frac{2}{1,000}$

Change these decimals to fractions and reduce when possible.

1. 0.5 = _____
2. 0.2 = _____
3. 0.253 = _____
4. 0.06 = _____
5. 0.023 = _____
6. 0.25 = _____

Fill in the equal values.

	Fraction	Decimal	Percent
1.	_____	0.18	_____
2.	$\frac{1}{4}$	_____	_____
3.	_____	_____	7%
4.	_____	0.95	_____
5.	$\frac{3}{10}$	_____	_____
6.	_____	_____	23%
7.	_____	_____	$6\frac{1}{2}\%$
8.	$\frac{1}{10}$	_____	_____
9.	_____	0.15	_____
10.	$\frac{1}{5}$	_____	_____

UNIT 1 ■ THE CALCULATOR AND A REVIEW OF THE BASICS

11. What is the value of the *shaded* area as a:

    A. Fraction _____

    B. Decimal _____

    C. Percent _____

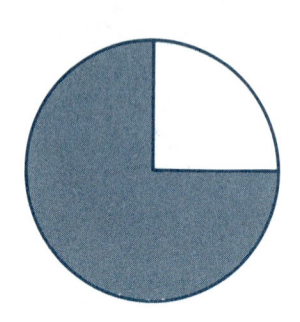

Changing Percents with Fractions to Decimals

To change a percent with a fraction to a decimal, change the fraction to a decimal before you change the percent to a decimal.

Example 1

$6\frac{1}{2}\%$ $2\overline{)1.00}$ with quotient 0.50

$6\frac{1}{2}\% = 6.50\% = 6.5\% = 0.065$

Example 2

$\frac{3}{4}\%$ $4\overline{)3.00}$ with quotient 0.75

$\frac{3}{4}\% = 0.75\% = 0.0075$

Change these percents with fractions to decimals.

1. $6\frac{1}{2}\% =$ _____

2. $18\frac{3}{4}\% =$ _____

3. $12\frac{1}{4}\% =$ _____

4. $5\frac{1}{5}\% =$ _____

5. $7\frac{1}{10}\% =$ _____

6. $36\frac{9}{10}\% =$ _____

7. $\frac{1}{2}\% =$ _____

8. $\frac{3}{4}\% =$ _____

RATIO AND PROPORTION

In addition to operations with fractions, decimals, and percents, you need to understand ratio and proportion to solve many problems.

Ratio

A ratio is a comparison of two numbers. You can write ratios in three ways:

$\frac{2}{3}$ or   2 to 3   or   2:3

Example In a family of 5 children there are 2 girls and 3 boys.

We could write the ratio of girls to boys as $\frac{2}{3}$.

We could write the ratio of boys to girls as $\frac{3}{2}$.

We could write the ratio of boys to the total number of children as $\frac{3}{5}$.

HINT: The number after the word *to* becomes the denominator when the ratio is written as a fraction.

Show the following information as a ratio. Write the ratio in three different ways and reduce the ratio if possible. The first problem is completed as an example.

1. The Washington Runners won 16 of their 24 games last season. What is the ratio of the games won to the total number of games played?

   A. $\frac{16}{24} = \frac{2}{3}$ B. 16 to 24 = 2 to 3   C. 16:24 = 2:3

2. In the school parking lot there were 20 student cars. Seventeen of the cars were low riders. Show the ratio of the total cars to the low-riders cars.

   A. _____   B. _____   C. _____

3. Six out of 10 households watched the Academy Awards on TV. Show the ratio of those who watched the Awards to the total number of households.

   A. _____   B. _____   C. _____

4. The homemaking class has 18 students; 10 of the students are boys. Show the ratio of the number of boys in the class to the number of girls in the class.

   A. _____   B. _____   C. _____

UNIT 1 ■ THE CALCULATOR AND A REVIEW OF THE BASICS   41

Proportion

A proportion is two equal ratios, which can be written in the following ways:

$$\frac{2}{4} = \frac{1}{2} \qquad 2 \text{ to } 4 = 1 \text{ to } 2 \qquad 2:4 = 1:2$$

Example Two of every four records are purchased by teenagers. This is the same ratio as one of every two records.

2 of 4 = 1 of 2

By cross multiplying, we can prove that the two ratios in a proportion are equal.

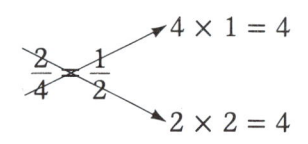

$$\frac{2}{4} \times \frac{1}{2} \qquad \begin{array}{l} 4 \times 1 = 4 \\ 2 \times 2 = 4 \end{array}$$

Decide which of the following are proportions and circle them.

1. $\frac{50}{1} = \frac{2}{1}$

2. $9:1 = 18:2$

3. $5 \text{ to } 15 = 2 \text{ to } 3$

4. $3:6 = 1:2$

5. $\frac{3}{8} = \frac{10}{24}$

6. $100 \text{ to } 50 = 1 \text{ to } 2$

Finding a Missing Number in a Proportion

The examples below show how to find the missing number (represented by a ?) in a proportion.

Example 1

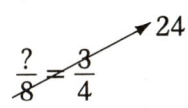

$$\frac{?}{8} = \frac{3}{4}$$

Cross multiply: $8 \times 3 = 24$

Divide the remaining number (4) into the cross multiplying result (24).

$$\begin{array}{r} 6 \\ 4\overline{)24} \end{array} = \begin{array}{l}\text{the missing number} \\ \text{cross multiplication result}\end{array}$$

HINT: Check your answer by cross multiplication. The answers should be the same.

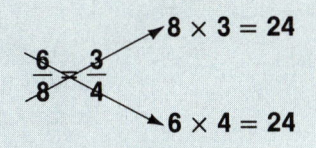

$8 \times 3 = 24$
$6 \times 4 = 24$

Find the missing number in each of the following proportions. Check your answer by cross multiplying.

1. $\dfrac{2}{8} = \dfrac{?}{16}$

2. $\dfrac{1}{6} = \dfrac{3}{?}$

3. $5:8 = ?:16$

4. ? to 12 = 1:3

5. $\dfrac{4}{?} = \dfrac{3}{12}$

6. 5 to 6 = 10 to ?

7. $8:? = 2:4$

8. $\dfrac{6}{4} = \dfrac{18}{?}$

9. $\dfrac{10}{100} = \dfrac{1}{?}$

UNIT 1 ■ THE CALCULATOR AND A REVIEW OF THE BASICS

Many problems can be solved by writing the information as a proportion. It is very important to label each part of the proportion. The following example shows the steps in solving a problem using a proportion.

Example 2   Mrs. Tanaka is driving to San Francisco. If she can drive 424 miles in 8 hours, how many miles can she drive in 14 hours?

Step 1

Choose two numbers that are related. Write them as a ratio. Label them.

$$\frac{8 \text{ hours}}{424 \text{ miles}}$$

Step 2

Set up a fraction for the second ratio. Label the second ratio as you did the first.

$$\frac{\text{hours}}{\text{miles}}$$

Step 3

Complete the second ratio using the known value and a question mark for the unknown value.

$$\frac{8 \text{ hours}}{424 \text{ miles}} = \frac{14 \text{ hours}}{? \text{ miles}}$$

Step 4

Solve the problem by cross multiplying.

$$\frac{8 \text{ hours}}{424 \text{ miles}} \nearrow \frac{14 \text{ hours}}{? \text{ miles}} \quad 5{,}936$$

$$5{,}936 \div 8 = 742$$

Answer: 742 miles

HINT: Note that hours and hours are in the numerators of both ratios; miles and miles are in the denominators of both ratios.

Write the following as proportions. Label all information.

1. On Friday night Sam cruised Main Street for 2 hours. He drove 5 miles. How many miles per hour did he average? _____

2. If Janet earns $68 in 4 weeks, how many weeks will it take her to earn $102? _____

Write the following as proportions, *label* them, and *solve*, with or without using the calculator. The first one has been done for you.

1. A store sold 136 sets of stoneware dishes for a total amount of $5,372. How much would 50 sets cost?   *$1,975*

$$\frac{136 \text{ sets}}{\$5{,}372 \text{ cost}} \nearrow \frac{50 \text{ sets}}{? \text{ cost}} \quad \$268{,}600$$

$$\$268{,}600 \div 136 = \$1{,}975$$

4. Betty Proper is having a party and plans to serve stuffed peppers. She paid $4.90 for 5 pounds of green peppers. How much did the peppers cost per pound?

_____

3. A sidewalk poll showed that out of 30 people questioned, 25 favored handgun control. If 42 people were questioned, how many would favor handgun control?

_____

4. A city used 16,200 bricks on the pedestrian crosswalks on Willow Pass Road, covering 1,800 square feet. How many bricks will the city need to cover 300 square feet on the crosswalks on Salvio Street?

_____

5. 660,000 people live in Contra Costa County. Four tenths of the people moved to the county within the last 5 years. How many people moved to Contra Costa within the last 5 years?

_____

6. A local ambulance traveled 15 miles in 10 minutes. How many miles per hour did the ambulance average?

_____

7. An advertisement claims that 2 of every 3 doctors recommend taking vitamin C to relieve cold symptoms. If there are 180,000 doctors in the U.S., how many doctors recommend vitamin C? _____

8. If it takes 10 people 6 hours to assemble 100 car stereo kits, how long will it take to put 500 kits together? _____

PERCENT

Percent means per 100. Percent is used when something has been divided into 100 parts.

Examples 28% means 28 of 100, or $\frac{28}{100}$ or 0.28.
100% means the whole amount, or 1.
200% means twice the original amount, or 2.

One method of solving percent problems is to rewrite the problem showing that the percent over 100 is equal to the part over the total.

Example 1 $20 is 4% of what number?

Step 1

Rewrite the problem.

$$\frac{\%}{100} = \frac{\text{part}}{\text{total}}$$

$$\frac{4}{100} = \frac{\$20}{?}$$

Step 2

Cross multiply.

$$\frac{4}{100} = \frac{\$20}{?} \rightarrow \$2,000$$

$100 \times \$20 = \$2,000$

Step 3

Divide the result of cross multiplying by the remaining number.

$$4\overline{)\$2,000}^{\$500}$$

Answer: $20 is 4% of $500.

46   UNIT 1 ■ THE CALCULATOR AND A REVIEW OF THE BASICS

Example 2 What is 4% of $500?

Step 1

Rewrite the problem.

$$\frac{\%}{100} = \frac{\text{part}}{\text{total}}$$

$$\frac{4}{100} = \frac{?}{\$500}$$

Step 2

Cross multiply.

$$\frac{4}{100} = \frac{?}{\$500} \rightarrow \$2,000$$

$4 \times \$500 = \$2,000$

Step 3

Divide the result of cross multiplying by the remaining number.

$$100 \overline{) \$2,000}^{\$20}$$

Answer: 4% of $500 is $20.

Example 3 $20 is what percent of $500?

Step 1

Rewrite the problem.

$$\frac{\%}{100} = \frac{\text{part}}{\text{total}}$$

$$\frac{?\%}{100} = \frac{\$20}{\$500}$$

Step 2

Cross multiply.

$$\frac{?\%}{100} = \frac{\$20}{\$500} \rightarrow \$2,000$$

$100 \times \$20 = \$2,000$

Step 3

Divide the result of cross multiplying by the remaining number.

$$\$500 \overline{) \$2,000}^{4}$$

Answer: $20 is 4% of $500.

Solve the following problems with or without using your calculator.

1. Monica and Victoria played 50 sets of tennis. If Monica won 35 sets, what percent did she win? _____

2. Twelve percent of the local high school student body holds part-time jobs. If 135 students work, what is the total school population? _____

3. An airline is offering a 15% discount for plane tickets purchased in the month of February. What would a round-trip ticket to Washington, DC cost, if the original price is $339?

4. A fabric store is having a 30% sale on corduroy. How much will 10 yards of the corduroy cost if it regularly sells for $6.50 a yard?

5. In the first half of the school year, there were 90 school days. If the principal was absent 9 days, what percent of the school days did she miss?

6. A credit union pays $5\frac{1}{2}$% interest per year. How much interest will a savings account of $2,160 earn in 1 year?

7. A wool skirt sold for $23.80. The store's profit was $5.95. What percent was the store's profit? _____

8. The Dean weighs 190 pounds after dieting for 6 months. This is 80% of the Dean's previous weight. How much did the Dean weigh before dieting? _____

9. Ann and Carmen are saving to rent an apartment together. They need $950 for the first and last month's rent and $310 for a cleaning deposit. Their combined take-home pay is $1,400 a month. If they save 30% of their combined paychecks, how many months do they need to save to rent the apartment? _____

10. Andre's total sales for the day were $600. His cash was short by 4%. How much was he short? _____

UNIT 1 ■ THE CALCULATOR AND A REVIEW OF THE BASICS

11. The senior class set a goal to sell 1,000 magazines during the spring sale. During the first week the seniors sold 350 magazines. What percent of their goal did they achieve? _____

12. A sales clerk had a 15% gain in sales. The gain was $300. What was the clerk's total sales? _____

OTHER APPROACHES TO SOLVING PERCENT PROBLEMS

The following pages show three other ways to solve percent problems. If you are comfortable with the "percent over 100 is equal to the part over the total" method shown on page 46 and page 47, continue to use that method when solving the problems on the following pages.

Finding the Percent of a Number

To find the percent of a number, change the percent to a decimal and multiply.

Example 1

What is 40% of 250?

40% = 0.40

0.40 × 250 = 100

Answer: 40% of 250 is 100.

Example 2

What is 15% of $72?

15% = 0.15

0.15 × $72 = $10.80

Answer: 15% of $72 is $10.80.

Find the percent of the following numbers. Show your work.

1. 25% of $250 = _____   2. 75% of $150 = _____

3. 60% of $2,000 = _____   4. 30% of $24.50 = _____

There are two methods for finding the percent of a number on a calculator. If the calculator has a percent key, multiply the number by the percent.

Example 3   What is 25% of 75?

75 ☒ 25 ☒ 18.75

The answer, 18.75, will appear in the display *without* pressing the equals key.

If the calculator does not have a percent key, multiply the number times the decimal form of the percent.

Example 4   What is 6% of 35?

35 ☒ 0.06 ☒ 2.1

The answer, 2.1, appears in the display *after* pressing the equals key.

Use your calculator to solve the following percent problems.

1. 8% of $120 = _____   2. 10% of $80,000 = _____

3. 40% of $45.50 = _____   4. $6\frac{1}{2}$% of $100 = _____

5. The commuter trains have been running late, causing 30% of the 650 passengers to be late to work one morning. How many people were late?   _____

6. Don was in an automobile accident; as a result, the cost of his auto insurance went up 15%. The original annual payment was $750. How much did the company raise his insurance bill?   _____

UNIT 1 ■ THE CALCULATOR AND A REVIEW OF THE BASICS

7. Using the information in Problem 6, what is Don's *total* insurance premium now? _____

8. Sales tax in a county is 6.5%. How much will the tax be on $3.59? (Round to the nearest cent.) _____

Finding What Percent One Number is of Another

To find what percent one number is of another, rewrite the problem as a fraction. Then change the fraction to a percent.

Example 1   18 is what percent of 90?

Step 1

Rewrite the problem as a fraction. To help you determine which is the numerator and which is the denominator, think of part and whole: 18 is what part of 90?

$\dfrac{18}{90}$

Step 2

Divide the denominator into the numerator.

$$\begin{array}{r} 0.20 \\ 90\overline{)18.00} \\ \underline{180} \\ 0 \end{array}$$

Step 3

Change the decimal to a percent.

$0.20 = 20\%$

Answer: 18 is 20% of 90.

HINT: You can reduce the fraction to make the division easier. The answer will be the same.

$\dfrac{18}{90} = \dfrac{2}{10} = \dfrac{1}{5}$

$$\begin{array}{r} 0.20 = 20\% \\ 5\overline{)1.00} \\ \underline{10} \\ 0 \end{array}$$

Find what percent one number is of another. Show your work.

1. $18 is what percent of $20? _____

2. $45 is what percent of $450? _____

3. $25 is what percent of $500? _____

To find what percent one number is of another with your calculator, divide the total into the part. Change the decimal answer that appears in the display to a percent.

Example 2   6 is what percent of 20?

6 ÷ 20 = 0.3 = 30%

Solve the following problems using your calculator.

1. $50 is what percent of $250?

2. 8 is what percent of 10?

3. $20 is what percent of $2,000?

4. What percent of 200 is 10?

5. 216 is what percent of 720?

6. What percent of 810 is 486?

7. 153 is what percent of 900?

8. Tom had 96 marshmallows and ate 12 of them. What percent of the marshmallows did he eat?

9. After the second wash/rinse cycle of the dishwasher, 7 of the 35 milk glasses were water-spotted. What percent of the glasses were spotted?

10. Dan Velvet will receive a $70 bonus. What percent of his $175 weekly salary is this bonus?

Finding a Number When a Percent of the Number is Known

To find a number when a percent of the number is known, change the percent to a decimal and divide the decimal into the known number.

Example 1   10 is 40% of what number?

Step 1

Change the percent to a decimal.

40% = 0.40

Step 2

Divide.

$$0.40 \overline{)10.00} = 25$$

$$\begin{array}{r} 25 \\ 0.40\overline{)10.00} \\ \underline{8\,0} \\ 200 \\ \underline{200} \\ 0 \end{array}$$

Answer: 10 is 40% of 25.

Solve the following problems. Show your work.

1. 10 is 10% of what number?

2. 21 is 35% of what number?

3. 99 is 15% of what number?

There are two calculator methods for finding a number when the percent of the number is known. If the calculator has a percent key, divide the known number by the percent.

Example 2   25% of what number is 5?

5 ÷ 25 % 20

The answer, 20, will appear in the display *without* pressing the equals key.

If the calculator does not have a percent key, change the percent to its decimal equivalent. Divide the known number by the decimal.

Example 3 $6 is 8% of what amount?

6 ÷ 0.08 = 75

The answer, 75, appears in the display *after* pressing the equals key.

Use your calculator to solve the following problems.

1. $3 is 5% of what amount?

2. $3.72 is 15% of what amount?

3. 375 is 75% of what number?

4. Emma made cookies and ate 20 of them, or 25% of the total. How many did she make altogether?

5. A record store sells tapes at a 40% savings, a savings of $2.40 each. What is the original price of a tape?

6. How much money must be invested at 8% interest to earn $1,000 per year?

7. Pat ate 75% of a pizza, or 6 pieces. Into how many pieces was the pizza cut?

UNIT 1 ■ THE CALCULATOR AND A REVIEW OF THE BASICS

Solve the following problems with or without using your calculator.

1. The cost of living has gone up 9% this year. If your salary was $11,432 this year, how much do you need to make *next year* to keep up with the cost of living? _____

2. Sales tax is $6\frac{1}{2}$%. How much tax would you pay on a $350 television? _____

3. The tuxedo in the advertisement may be purchased for $151.25 and has a comparable value of $275. What percent could you *save* by buying the tuxedo for $151.25? _____

Tuxedos

Classic elegance for those special occasions. Fashioned from wool blend with full attention to detail.

$151.25

Compare at $275

4. Mr. Chang sold $505 worth of goods. His commission was 12%. How much did he earn? _____

5. The Rec Center put 200 sodas in the vending machine. If 80% were sold the first day, how many sodas were sold? _____

6. If each soda in Problem 5 cost 50¢, how much money was put into the vending machine that day? _____

7. A football team won 12 games, or 80% of the total number of games played. How many games did the team play? _____

8. Jason and Ramon's apartment will cost $475 a month to rent. What percent of their $1,350 take home pay will they pay in rent? (Round to the nearest percent.) _____

9. The National Academy of Sciences made a study of trash. The study showed that trash consists of the following items:

   60% paper
   16% cans
   6% bottles
   6% plastic products
   12% miscellaneous items

   The average American produces about 1,800 pounds of trash a year. Use the above percentages to calculate how many pounds one person produces of each of the following:

   A. paper _____
   B. cans _____
   C. bottles _____
   D. plastic _____
   E. miscellaneous items _____

UNIT 1 ■ THE CALCULATOR AND A REVIEW OF THE BASICS

10. Thirty percent of the students in a local high school have cars. If there are 450 students, how many students have cars? _____

11. Sean works at a department store and receives a 12% commission on his sales. Last month he sold $1,500 worth of goods. What was the total dollar amount of his commission for the month? _____

12. Of the 140 fires in an area this year, 84 were caused by children under 12 years of age. What percent of the fires were caused by children under 12? _____

13. A car battery regularly sells for $55.50. How much can you save if it is marked down 20%? _____

2 WORD PROBLEMS AND PROBLEM SOLVING

SKILLS YOU WILL NEED

Basic math skills
Willingness to try

SKILLS YOU WILL LEARN

How to approach a problem-solving situation
A variety of problem-solving strategies
An openness to solving problems more than one way

SITUATIONS IN WHICH YOU'LL USE THESE SKILLS

At home
On the job

Hear Everything   See Everything   Ask Questions   Think Some More

    Problem solving requires more than math. It requires confronting a problem with all your mental and physical resources. This problem-solving chapter is designed to help you develop strategies that can be applied to a variety of problems and situations.

    Being puzzled or perplexed is a natural state in problem solving. It is important to have a positive attitude towards something new, a willingness to accept that your first try might not work, and a realization that there is more than one way to solve a problem.

    After applying all the strategies discussed in this chapter, the solution might still escape you. It is then time to think some more. You may need to try another approach, perhaps one that uses algebra or geometry. You may need to visualize the situation or talk to others about the problem. You might need to ask yourself, "Have I solved a similar problem before?" When all else fails, you may need to sleep on it. Remember to keep an open mind towards new and unusual situations.

The following list suggests ways to approach a problem-solving situation. You probably won't need all the strategies to solve every problem you face. The first 10 strategies are discussed in this unit. The last 5 strategies are for you to consider if you haven't been able to solve a problem.

1. Follow directions.
2. Read problem carefully and look for key words.
3. Decide which math operation to use.
4. Estimate.
5. Simplify the problem and substitute easier numbers.
6. Look for likenesses and patterns.
7. Guess and check.
8. Draw a picture or diagram.
9. Manipulate (touch and move) objects.
10. Keep track of clues and information.
11. Try to think of other strategies.
12. Visualize.
13. Talk to others.
14. Think of a similar problem you have solved.
15. Sleep on it.

STRATEGY 1—FOLLOW DIRECTIONS

To solve a problem, it is very important that you follow whatever directions are given. This means that before you start working, you must read all the directions. Make sure you understand exactly what you are to do before you begin.

Example Read all of the following information.

Jim Neel received his weekly paycheck Friday. His gross pay was $328.67. From this amount, $63.27 was deducted for federal income tax, $18.64 was deducted for state income tax, and $24.14 was deducted for F.I.C.A.

He deposited his paycheck in his checking account, which had a balance of $33.55. That evening he wrote a check to a utility company for $45.73, and one to a telephone company for $74.86.

Jim is a careful consumer and wants to get the best value for his money. He needs to fill up his pickup truck with 15 gallons of gasoline, and must decide whether to buy from Horrible Harry, who advertises regular unleaded for $0.38 per liter, or from Lighthouse, who advertises regular unleaded for $1.39 per gallon.

You have read a lot of information about Jim Neel. There are many different problems that could be solved with this information, but, for now, do nothing with it. Congratulate yourself on reading all the directions and information before trying to solve a problem.

Follow the directions and answer the questions.

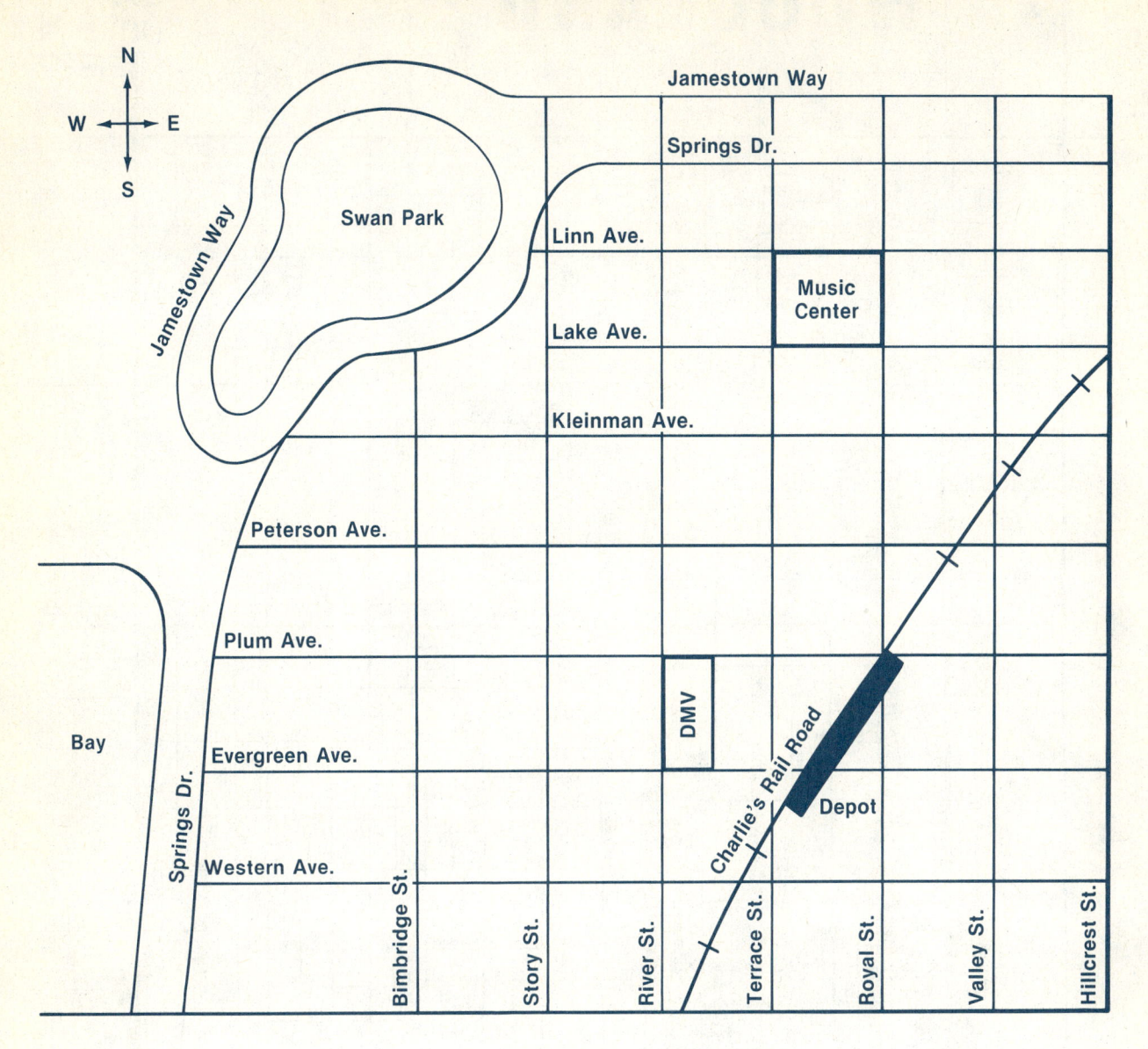

1. Using the map above, drive west on Jamestown Way, enter Kleinman Ave. and drive 3 blocks east. Make a turn north and continue 2 blocks. Turn right and you will be traveling towards: _____
    A. Swan Park
    B. The Music Center
    C. DMV Offices
    D. Charlie's Rail Road

2. From the DMV office on Plum on the map above, drive 1 block north; turn east and drive 1 block; head north 2 more blocks. Turn west. If you park there you will be on: _____
    A. Springs Drive
    B. Linn Ave.
    C. Lake Ave.
    D. Evergreen Ave.

UNIT 2 ■ WORD PROBLEMS AND PROBLEM SOLVING

3. Using the map above, enter Woodland on Highway 113, traveling north. Turn left 1 block north of the Greyhound Depot, travel 7 blocks, and turn left again. Travel 3 blocks and turn west. Travel 3 blocks and park your truck. You are parked near: _____
   A. Dept. Motor Vehicles
   B. City Park
   C. Central Place
   D. Library
   E. Yuba Community College

4. You are to continue on your trip to Clear Lake, still using the map above. Clear Lake is west of Woodland. On which highway should you be traveling to reach Clear Lake? _____

UNIT 2 ■ WORD PROBLEMS AND PROBLEM SOLVING

5. A. Draw the letter O with a 1-inch radius centered around the dot below.
   B. Draw a $2\frac{1}{2}$-inch capital I with $\frac{1}{2}$-inch cross bars, lying on its side, touching the center top of the O, and extending equally in both directions.
   C. Draw the letter W, 1 inch tall and approximately 1 inch in width; make the top two points touch the bottom of the O and center them below the dot.
   D. Draw a $1\frac{1}{2}$-inch-long letter V, with the top two points of the V touching the bottom points of the letter W.
   E. On the left, outside the letter O, draw a letter C approximately 1 inch tall; center it opposite the dot with both points touching the letter O.
   F. On the right side of the letter O, draw a matching backward 1-inch letter C.
   G. Draw the letter A, $1\frac{1}{2}$ inches wide and $1\frac{1}{4}$ inches tall, with the bottom points touching and centered on the capital I.
   H. Draw two $\frac{1}{4}$-inch-diameter Os, resting on an imaginary line that extends through the dot and Cs. The Os need to be $\frac{1}{4}$ inch from the center dot and on opposite sides of it.
   I. Draw the bottom half of a U, $\frac{3}{4}$ inch wide. It should be $\frac{1}{4}$ inch below the small Os.
   J. What have you drawn? _____

●

UNIT 2 ■ WORD PROBLEMS AND PROBLEM SOLVING

STRATEGY 2—READ PROBLEM CAREFULLY AND LOOK FOR KEY WORDS

In order to solve a problem, you must understand what is being asked. Key words and information help you to understand what is being asked and how to solve the problem. In the example below, the key words or numbers have been underlined, the question has been restated, and the steps of the problem have been written out.

"This is a lot of work, but he did say to count the olives..."

Example Gloria bought a car for $9,480. It decreased in value by 25% the first year she owned it. What dollar amount did it decrease?

A. Restated Question

What is 25% of $9,480?

B. Steps of Problem

```
   $ 9,4 8 0    cost
 ×      0.2 5   decrease
   4 7 4 0 0
   1 8 9 6 0
   $ 2,3 7 0.0 0   decrease
```

C. Answer

$2,370

HINT: Sometimes you need to restate more than one question.

Underline the key words in the following problems. Restate the question, identify the steps, and solve.

1. A carpenter glued two pieces of wood together to make a breadboard. One piece of wood was $\frac{5}{16}$ inch thick and the other was $\frac{7}{8}$ inch thick. What is the thickness of the breadboard?

   ### A. Restated Question

   _____

   ### B. Steps of Problem

   ### C. Answer

   _____

2. If one shelf in a supermarket warehouse will hold 84 cans of fruit, how many shelves will be needed for 3,024 cans?

   A. Restated Question
   _____

   B. Steps of Problem   C. Answer

                                          _____

3. Gordon owes 6 payments of $26.35 each to a department store and 3 payments of $15.37 each to an oil company. How much greater is his first bill than his second bill?

   A. Restated Question
   _____
   _____

   B. Steps of Problem   C. Answer

                                          _____

4. In January, David decides to save money to buy a portable tape deck and headset. He plans to set aside $10 a week toward the $80 he will need. If he begins saving money on January 14 and continues saving the money regularly, in what month can he buy his tape deck and headset?

   A. Restated Question
   _____
   _____

   B. Steps of Problem   C. Answer

                                          _____
                                          _____

UNIT 2 ■ WORD PROBLEMS AND PROBLEM SOLVING

5. Sara has a choice of buying a coat for cash for $138 or paying $10 down with 4 monthly payments of $35 each. How much will she save if she pays cash?

   A. Restated Question
   _____
   _____

   B. Steps of Problem   C. Answer
                                      _____

Read each of the following problems carefully and solve.

1. A can of orange juice weighs 8 ounces. If there are 48 cans of juice in a case, and the empty case weighs 28 ounces, what is the total weight of the case of orange juice?   _____

2. Bonnie bought a dozen watches for $145. She sold the watches for $18.95 each. How much profit did she make on the dozen watches?   _____

3. $1 \times 2 \times 3 \times 4 \times 5 \times 6 \times 7 \times 8 \times 9 \times 0 =$

4. Which of the following will contain the most liquid?
   - A. One $2\frac{1}{2}$-gallon jar
   - B. One $\frac{1}{2}$-gallon jug
   - C. One 5-pint pail
   - D. One 3-liter bottle

   | 1 gallon = 128 ounces |
   | 1 liter = 33.8 ounces |
   | 1 quart = 32 ounces |
   | 1 pint = 16 ounces |

5. A student was standing on the middle stair talking to his friends between classes. When he heard the bell, he went up 4 steps, dropped his pencil, and had to go back down 7 steps to get it. He walked up 5 steps and met a friend going down. He had to tell the friend something, so he walked back down 9 stairs *to the bottom of the stairs*. When the tardy bell rang, he ran up the stairs to the top. How many stairs did he run up?

UNIT 2 ■ WORD PROBLEMS AND PROBLEM SOLVING

STRATEGY 3—DECIDE WHICH MATH OPERATION TO USE

In solving problems, it is very important to decide which math operation to use. Sometimes it is helpful to ask yourself, "Should I add, subtract, multiply or divide?" The following problem might seem confusing until you decide which math operation to use.

Example An antique dealer bought a rocking chair for $40 and sold it for $80. The dealer then bought it back for $120 and sold it again for $160. Did the dealer gain, lose, or break even? If the dealer gained or lost, what was the amount?

Subtract:

$80	sold
− 40	bought
$40	profit from first sale

Subtract:

$160	sold
− 120	bought
$ 40	profit from second sale

Add:

$40	profit from first sale
+ 40	profit from second sale
$80	total profit

Answer: The dealer gained $80.

Decide which math operation to use and solve the following problems.

1. Dick France has been working for 8 months and has earned 10 days of vacation. How many days of vacation does he earn each month?

2. A new rock group called Charged Up is giving a concert to benefit a local charity. Lucy is selling tickets to the concert. For every 3 tickets she sells, the price of her ticket is reduced by $1. If a ticket sells for $10, how many tickets does Lucy have to sell to attend the concert for free?

3. A 21-pound turkey must cook 18 minutes a pound. At what time will the turkey be done if it is put in the oven at 1:30 P.M.? _____

4. Barry bought a motorcycle that cost $1,100. He made a $250 down payment and borrowed the rest from his credit union. He must pay $170 to the credit union in interest fees. How much will he repay the credit union per month, if he pays off his loan in 10 months? _____

5. Carole needed a set of spark plugs for her car. How much would 8 spark plugs cost, if the price was 3 for $2.25? _____

UNIT 2 ■ WORD PROBLEMS AND PROBLEM SOLVING

STRATEGY 4—ESTIMATE

Estimating means guessing an answer before solving the problem. An estimated answer should be close to the actual answer, but the estimate and the actual answer probably won't be exactly the same. By estimating the answer, you have an idea of what the answer should be.

First decide how to solve the problem, then estimate the answer. Check your solution with your estimated answer.

Example 1

That was a good party. How much did it cost?

I spent $18.80 for food and $12.50 for fruit punch.

Four of us gave the party, so how much do we each owe?

I guess the party cost about $30 altogether, so if we multiply 4 times $30, that means we each owe $120.

That doesn't sound reasonable. I estimate that we each pay about $7. We need to find the total spent and divide by 4.

Solution

```
  $18.80    food              $   7.8 2   owed per person
+  12.50    fruit punch     4)$ 3 1.3 0
  $31.30    total spent        2 8
                                 3 3
                                 3 2
                                   1 0
                                      8
                                      2
```

Read the following problems, decide how to solve them, and circle the best estimated answers.

1. Before you go out Saturday, you have to wash the outside of the living room windows, wash the breakfast dishes, and change your clothes. Estimate how long it would take you to do these tasks.

   15 minutes $1\frac{1}{2}$ hours $5\frac{1}{2}$ hours

2. You are going to the store to buy a quart of milk, a loaf of bread, a dozen eggs, and $\frac{1}{2}$ dozen oranges. About how much money should you take?

    $1 $5 $10

3. Casey earned $85.65 last week and $105.50 this week at his after-school job. Estimate Casey's take-home pay if 20% was deducted from these amounts.

    $50 $100 $150 $200

Estimate answers for the following problems. Then solve the problems and compare your solutions to your estimates.

1. A tape deck costs $239.95 at a department store. The same tape deck costs $175.95 at a discount store. How much can be saved by shopping at the discount store?

    A. Estimate _____

    B. Solution _____

2. A package of 5 pairs of socks sells for $8.95. How much would each pair cost?

    A. Estimate _____

    B. Solution _____

3. Oscar earns $2.90 an hour. Last week he worked 32 hours. How much will he get paid for the week?

    A. Estimate _____

    B. Solution _____

UNIT 2 ■ WORD PROBLEMS AND PROBLEM SOLVING

The ability to estimate is extremely important when using the calculator, because it is very easy to push the wrong button or to enter the problem incorrectly.

Example 2   A package of spaghetti costs 85¢ and contains 5 servings. What is the cost per serving?

Estimate: $1.00 ÷ 5 = $0.20

Incorrect Entry: 5 [÷] 0.89 [=] 5.6179775

$5.62 is not close to the estimate and is not reasonable. The problem was entered into the calculator incorrectly.

Correct Entry: 0.85 [÷] 5 [=] 0.17

Use your calculator to help you complete the following table.

Problem	Estimated Problem	Estimated Answer	Calculator Answer
2,862 ÷ 54	3,000 ÷ 50	60	53
9,780 ÷ 962			
627.75 ÷ 15.2			
970.5 ÷ 5			
1,127.8 ÷ 13.15			

Read the following problems to determine whether the answer is reasonable and whether the problems have been correctly keyed on the calculator.

1. Dawn wants to buy a pair of boots for $59.79, a jacket for $73.65, and gloves for $7.95. How much money does she need altogether?

   [AC] 59.79 [+] 73.65 [+] 7.95 [=] 141.39

   A. Is the answer reasonable? _____

   B. If not, what is the error? _____

2. Juanita earns $82.75 a day. Last month she worked 21 days. How much did she earn?

   [AC] 82.75 [×] 12 [=] 993

   A. Is the answer reasonable? _____

   B. If not, what is the error? _____

UNIT 2 ■ WORD PROBLEMS AND PROBLEM SOLVING

3. Jack pays $15 a week towards a bicycle that costs $180. How many payments must he make?

   AC 15 ÷ 180 = 0.0833333

   A. Is the answer reasonable? _____

   B. If not, what is the error? _____

4. Jana is going to buy a car for $2,395. How much will she have to pay in sales tax if the tax rate is 6%?

   AC 2,395 × 0.6 = 1,437

   A. Is the answer reasonable? _____

   B. If not, what is the error? _____

5. A local department store charges 1.8% interest a month on any unpaid balance. If you have an unpaid balance of $360, how much finance charge will you pay next month?

   AC 360 ÷ 0.018 = 20,000

   A. Is the answer reasonable? _____

   B. If not, what is the error? _____

Estimate answers for the following problems. *Solve the problems with a calculator* and compare your solutions to the estimates.

1. Gary deposited his paycheck of $109.81 to his checking account, which had a balance of $15.24. If he writes a check to Auto-Matic for $39.16 and a check to PayLow for $21.68, can he make his $80 car payment before next payday?

   Estimate (yes or no) _____

   Solution   A. Has _____

   B. Needs _____

2. You just received $50 from your grandmother for your birthday. If you want to use all of your gift at your favorite record shop, how many tapes at $7.89 each could you buy?

   Estimate _____

   Solution _____

3. Rose is paying $120 a month for her car payment. If she will make a total of 18 payments, how much will the car cost?

   Estimate _____

   Solution _____

UNIT 2 ■ WORD PROBLEMS AND PROBLEM SOLVING

4. A machine bolt weighs 45 grams. How much will a box of 215 bolts weigh?

Estimate _____

Solution _____

STRATEGY 5—SIMPLIFY THE PROBLEM AND SUBSTITUTE EASIER NUMBERS

Some word problems are confusing, and you may not know whether to add, subtract, multiply, or divide. When in doubt, follow these steps:

1. Rewrite the problem, using easier numbers, and solve the simplified problem.
2. Decide what you did to solve the simplified problem (add subtract, multiply, or divide) and whether you used more than one step to solve the problem.
3. Solve the original problem by using the same steps you used to solve the simplified problem.

Example 1 If you buy a used car for $850 and pay $62 in interest charges, how much will your monthly payments be if you repay the loan over a period of 12 months?

Step 1

Rewrite the problem:
If I bought a car for $50 and paid $10 interest, how much would I repay each month, over a period of 12 months?

$50
+ 10
$60

$60 ÷ 12 = $5

Step 2

Decide what you did to solve the simplified problem:
Add the loan plus interest:

$50   loan
+ 10   interest
$60   total owed

Divide the number of months into the total amount owed:

$ 5 ← monthly payment
12) $60
   ↑   ↖
months   total owed

Step 3

Solve the original problem by using the same steps you used to solve the simplified problem.

$850   loan
+ 62   interest
$912   total owed

```
       $ 76   monthly payment
   12 ) $912
         84
         ---
         72
         72
         ---
          0
```

Answer: $76

74   UNIT 2 ■ WORD PROBLEMS AND PROBLEM SOLVING

> **HINT:** Easier numbers may be smaller numbers as in Example 1 or they may be rounded numbers.

Example 2 Jackie and Mike bought a couch for $589 and a matching recliner chair for $338. If they pay 18.5% interest on a furniture loan for 1 year, what will be the total cost of the furniture?

Step 1

Rewrite the problem:
They bought a couch for $600 and a chair for $300. They pay 20% interest for 1 year. What is the total cost?

$600 + $300 = $900

$900 × 0.20 = $180

$900 + $180 = $1,080

Step 2

Decide what you did to solve the simplified problem:
Add the cost of the couch plus the cost of the chair:

```
  $600    couch
+  300    chair
  $900    cost of furniture
```

Multiply the interest rate times the cost of the furniture:

```
    $ 9 0 0      cost of furniture
×     0.2 0     interest rate
  $ 1 8 0.0 0   amount of interest
```

Add the amount of interest and the cost of the furniture:

```
$  900    cost of furniture
+  180    interest
$1,080    total cost
```

Step 3

Solve the original problem by using the same steps you used to solve the simplified problem.

```
  $589    couch              $ 9 2 7       cost of furniture
+  338    chair          ×     0.1 8 5    interest rate
  $927    cost of furniture    4 6 3 5
                               7 4 1 6
                                 9 2 7
                           $ 1 7 1.4 9 5 = $171.50   amount of interest
```

```
$  927.00    cost of furniture
+  171.50    amount of interest
$1,098.50    total cost
```

Answer: $1,098.50

UNIT 2 ■ WORD PROBLEMS AND PROBLEM SOLVING

Substitute easier numbers in the following problems, and write those numbers in the spaces provided. In these problems use rounded numbers for the easier numbers. Solve and label the simplified problem.

1. Hank worked a 40-hour week at $4.25 _____ an hour. From his check was deducted $8.84 _____ for social security, $1.76 _____ for union dues, and $12.19 _____ for income tax. What was his take-home pay?

   _____

2. Margie is building a bookcase. The bookcase will have 6 wood shelves that attach to metal brackets on the wall. Each shelf will be $5\frac{1}{4}$ _____ feet long. How much lumber will she need to buy?

   _____

3. Which is a better buy, an 11 _____ -ounce can of ravioli at $1.39 _____ or a 6 _____ -ounce can at $0.47 _____ ?

   _____

4. Denim fabric is on sale and costs $29.92 _____ for 8 _____ yards. How much would $5\frac{1}{5}$ _____ yards cost?

   _____

76   UNIT 2 ■ WORD PROBLEMS AND PROBLEM SOLVING

Solve the following problems. When necessary, substitute easier numbers to help you decide how to solve the problem.

1. Barbara sold her car to her brother for $625. She received a $75 down payment. If her brother repays the balance over a period of 10 months (no interest charged), how much will Barbara receive each month? _____

2. Which is the best buy: green toothpaste at 98¢ for 13 ounces, blue toothpaste at 81¢ for 9 ounces, or white toothpaste at 59¢ for 6 ounces? _____

3. Frank borrowed his parents' credit card to shop for school clothes. He bought shoes for $19.95, 2 pairs of jeans for $15.85 each, 6 pairs of socks at 3 pairs for $4.50, and a sweat shirt for $8.99. If he spent $20 of his own money on the clothes, what amount did he charge on the credit card? _____

4. Carl's checking account balance was $109.63. He deposited the following three checks to his checking account: $18.75, $15.90, and $36.40. He then wrote a check for $89.46 to pay his gasoline bill. What is his new checkbook balance? _____

UNIT 2 ■ WORD PROBLEMS AND PROBLEM SOLVING

5. Lisa was backpacking in Europe and was shopping for a new, lightweight, 10-speed bike. The bike shop showed her a model that weighed 15 kilograms. If 1 kilogram is equal to 2.2 pounds, how much does the bike weigh in pounds?

_____

6. Yuki had trouble selling her 10-speed bicycle, so she reduced her asking price by 20%. If she wanted $80 originally, what is the new asking price?

_____

7. The Chavez family bought 65 kilograms of kitchen tile when they were in Mexico on vacation. If 1 kilogram is equal to 2.2 pounds, how many pounds of tile will they have shipped back to California?

_____

8. Phil is planning a barbecue and will buy 11 dozen hot dogs. If the hot dogs sell for $2.10 for 3 dozen, how much will 11 dozen cost?

_____

9. Danielle went to a pizza parlor to buy 9 salami pizzas. If the price was $7.50 for 2 pizzas, how much would 9 pizzas cost? _____

10. When buying a turkey weighing 12 pounds or less, one should allow $\frac{3}{4}$ pound per serving. How many people could be served from a 12-pound turkey?

    _____

11. Ralph can either borrow $250 from the credit union for $43 interest or pay the local bank 18.5% interest.

    A. Which is the better rate? _____

    B. How much can he save? _____

UNIT 2 ■ WORD PROBLEMS AND PROBLEM SOLVING

STRATEGY 6—LOOK FOR LIKENESSES AND PATTERNS

In order to solve some problems, it is necessary to look for likenesses and patterns or to visualize how things might work.

Example Below is a square with 4 holes. The square was folded in such a manner that a single punch produced this pattern. Look at the pattern and folds.

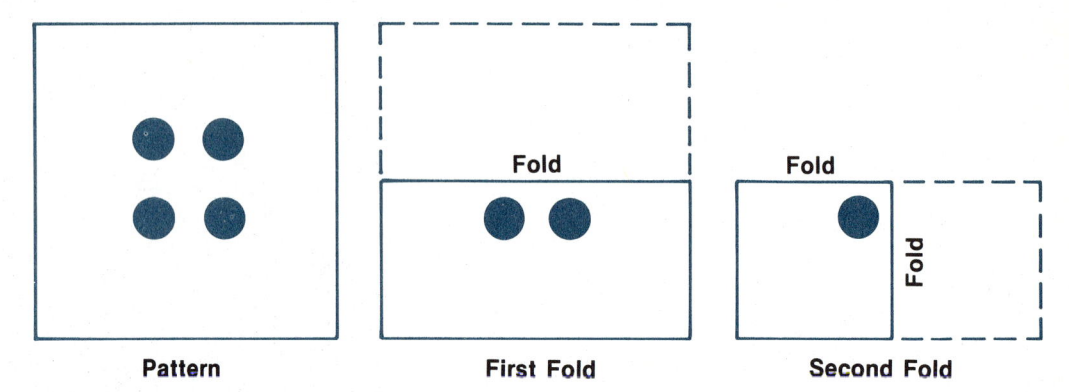

Pattern **First Fold** **Second Fold**

Look for likenesses and patterns to solve the following problems.

1. Cut out several small paper squares. Fold a square and make one punch to create the following patterns.

   A.    B. 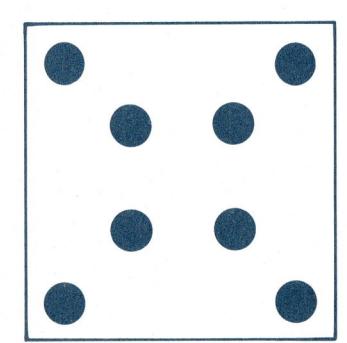   C.

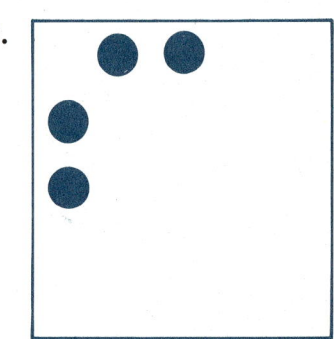

80 UNIT 2 ■ WORD PROBLEMS AND PROBLEM SOLVING

2. On Monday, Garo wrote a letter to the principal requesting better school lunches. On Tuesday, he asked 3 of his friends to write the principal and to contact 3 friends each to write letters on Wednesday. The Wednesday letter-writers each contacted 3 friends to write letters on Thursday. Each Thursday letter-writer contacted 3 friends to write letters on Friday. How many letters did the principal receive on Friday? _____

3. Five students were caught passing notes to each other. They each passed one note to each other. How many notes did the teacher collect from the 5 students? _____

UNIT 2 ■ WORD PROBLEMS AND PROBLEM SOLVING

4. The following poem tells the tale of Humphrey the whale. Look for likenesses and patterns to break the alphabet code. Each symbol represents a letter of the alphabet. The first two words are done for you. (Hint: Write the letters of the alphabet from A to Z. Each time you guess a symbol put the symbol by the letter.)

Humphrey ⌐∙∶○ =‖⋮ top ⟨∙⎮ =‖⋮ news ⩔⩔ 1985.

=‖⋮ country ⌐∙∶○ elated =‖⋮= ‖⋮ could survive.

While heading ○⟨⏋=‖ ‖⋮ ⊙⋮⟨∶⋀⋮ confused

And ⩔⩔=⟨ San Francisco Bay ‖⋮ ⟨∙⏋⩔○⋮⟩

∶⋀⟨⩔○○= =‖⋮ sails ∶⩔⟩ ships ∶⩔⟩ boats

Humphrey ⌐∙∶○ =‖⋮ most amazing spectacle ∶⎮∙▯⟨∶=

⏋⟩ =‖⋮ ⟩⋮▯=∶ ‖⋮ made his ⌐∙∶⌐

⟩⋮⟨⟩▯⋮ crowded =‖⋮ ⊙∶⩔▯○ =⟨ ⌐∙∶=⟨‖ him play.

=⟨ ○=⋮⋮∙⎮ him back towards the Golden Gate,

⩔⟨=‖⩔⩔○ ⌐∙⟨⏋▯⟩ ⟩⟨ ⊙⏋= =‖⋮ ○⟨⏋⩔⟩○ ⟨∙⎮ ∶ ⋀∶=⋮

5. How many squares do you see in the figure below? (Hint: There are more than 25.) _____

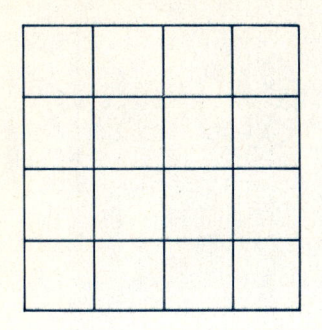

6. Ellen was packing clothes for a vacation. She packed 3 shirts, 3 blouses, and 3 pairs of pants.
   A. How many different outfits could she put together, wearing 1 top with 1 pair of pants? _____

   Each pair of pants can be worn with 6 different tops.
   3 × 6 = 18

   B. How many different outfits could she put together if the pants were worn with shirt and blouse combinations (the layered look)? _____

UNIT 2 ■ WORD PROBLEMS AND PROBLEM SOLVING

Decide which figures can be made by folding these patterns. Circle the correct letter.

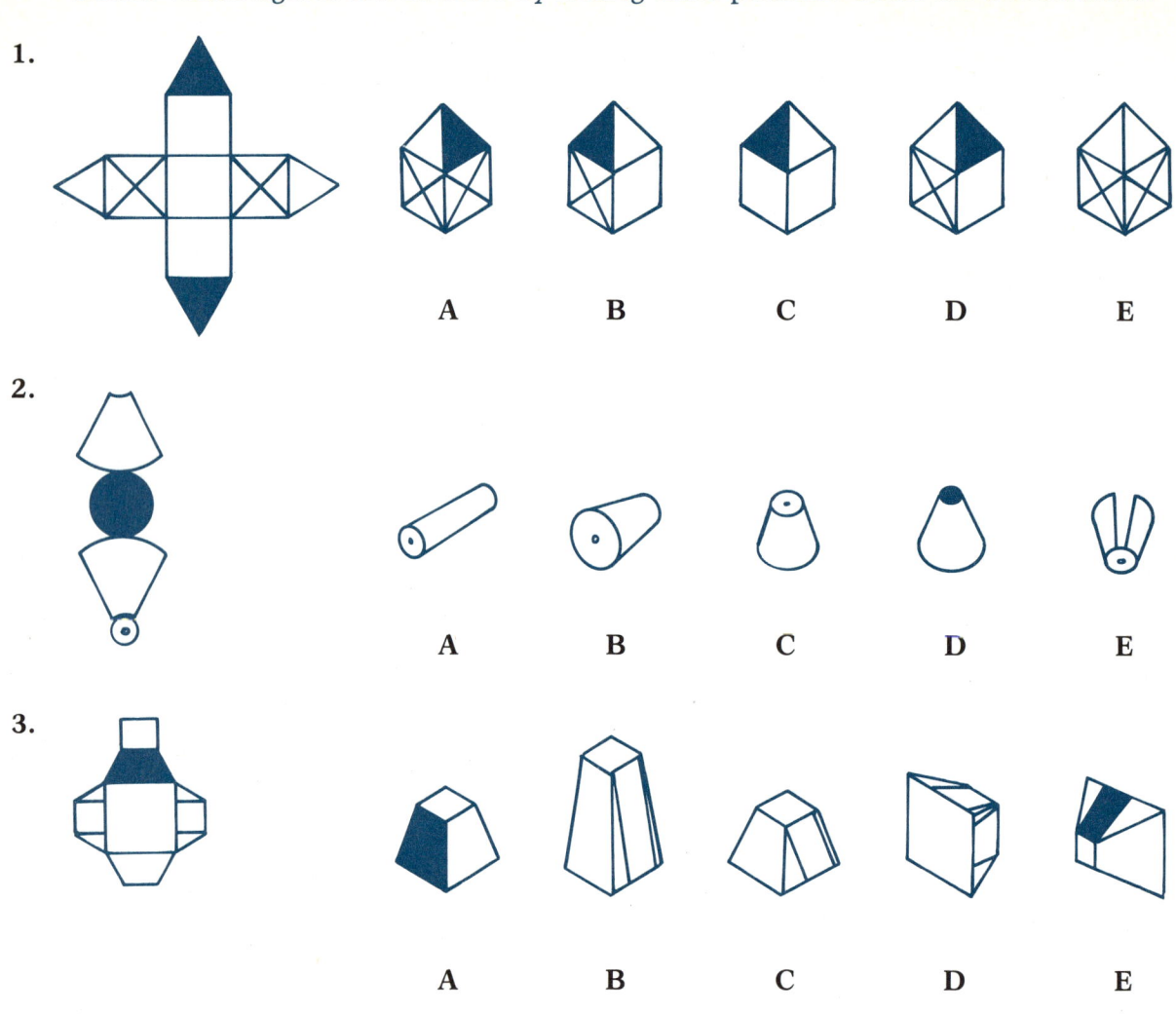

1.   A   B   C   D   E

2.   A   B   C   D   E

3.   A   B   C   D   E

Which of the four objects is like the object at the left? Circle the correct letter.

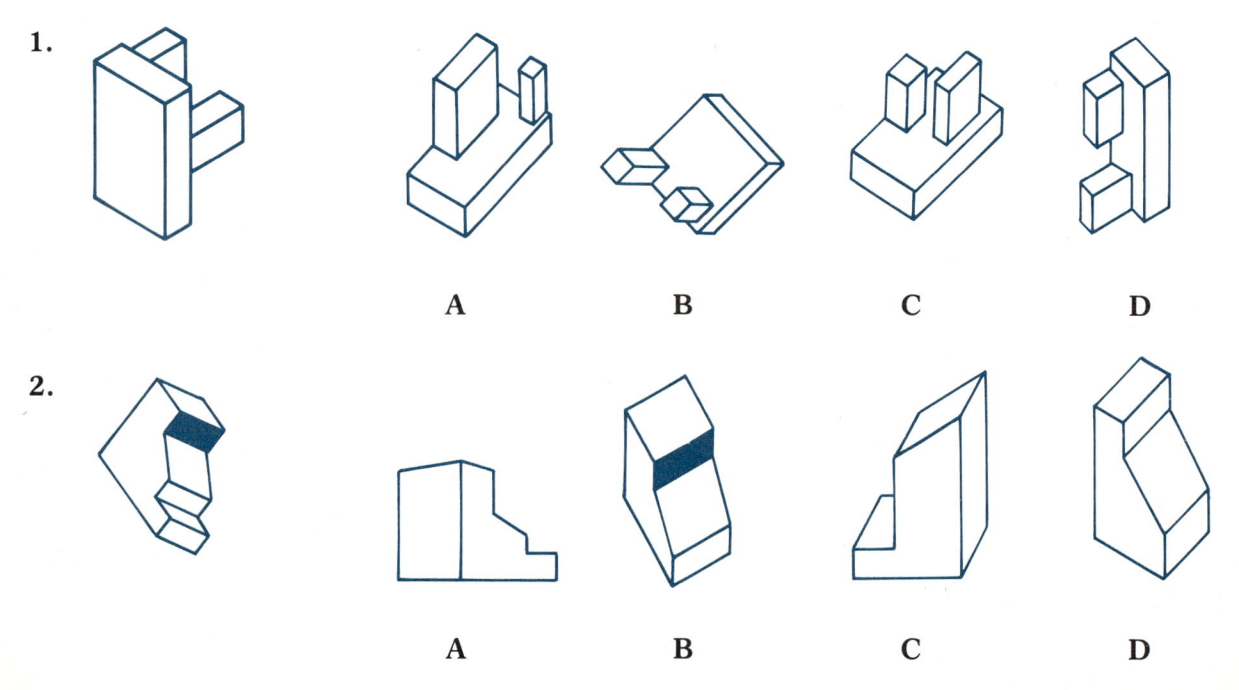

1.   A   B   C   D

2.   A   B   C   D

84   UNIT 2 ■ WORD PROBLEMS AND PROBLEM SOLVING

STRATEGY 7—GUESS AND CHECK

When you can't figure out how to solve a problem, sometimes the guess-and-check method will work. Simply guess an answer and check to see whether it makes sense. Based on your results, guess and check again. Keep using your results until you guess the right answer.

Example You must cut a 10-pound block of cheese into 4 pieces so that each piece is twice as heavy as the one before. How much will each piece weigh?

First Guess

  1 pound
  2 pounds
  4 pounds
  8 pounds
 ―――――――――
 15 pounds is too much

Second Guess

 $\frac{1}{2}$ pound
 1 pound
 2 pounds
 4 pounds
 ―――――――――
 $7\frac{1}{2}$ pounds is not enough

Third Guess

(The first piece must weigh between $\frac{1}{2}$ pound and 1 pound.)

 $\frac{3}{4}$ pound
 $1\frac{1}{2}$ pounds
 3 pounds
 6 pounds
 ―――――――――
 $11\frac{1}{4}$ pounds is too much

Fourth Guess

(The first piece must weigh between $\frac{1}{2}$ pound and $\frac{3}{4}$ pound.)

 $\frac{2}{3}$ pound
 $1\frac{1}{3}$ pounds
 $2\frac{2}{3}$ pounds
 $5\frac{1}{3}$ pounds
 ―――――――――
 $8\frac{6}{3}$ pounds = 10 pounds is the correct answer

Use the guess-and-check strategy to solve the following problems.

1. Spot just had a litter of puppies. Each female puppy has twice as many brothers as she has sisters. Each male puppy has the same number of sisters as he has brothers. How many male puppies and female puppies did Spot have? _____

UNIT 2 ■ WORD PROBLEMS AND PROBLEM SOLVING

2. Diana and Christopher are photographers. Last month their combined income was $9,000. Diana made $1,500 more than twice what Christopher made. How much did Christopher make? _____

3. Ms. Booth's creative writing class received $246 for selling 201 copies of its book of student poetry and short stories. If students were charged $1 for the book, and nonstudents were charged $2, how many nonstudents purchased the book? _____

4. Adam is $\frac{1}{6}$ as old as his mother. Added together, their ages total 28. How old is each? _____

5. Using the numbers 1 through 7, one time each, write an equation equaling 100. You may add, subtract, multiply, or divide in whatever manner you wish. _____

6. Mr. Dee, the driver's education teacher, told his class that he noticed 30 vehicles, which had a total of 80 tires, in the faculty parking lot.

   A. How many teachers parked their cars in the faculty parking lot?

   _____

   B. How many teachers parked their motorcycles in the faculty parking lot?

   _____

86 UNIT 2 ■ WORD PROBLEMS AND PROBLEM SOLVING

STRATEGY 8—DRAW A PICTURE OR DIAGRAM

Drawing a picture or diagram is very helpful in solving problems. The picture allows you to see the problem. Be sure to identify and label the parts of your diagram.

Example   The quarterback for the pirates threw a successful pass from his own 16-yard line to the Cougars' 25-yard line. How many yards did he throw the ball?

Below is a picture of the football field, the yardage markings, and the ball positions. This illustration shows that he threw the ball 59 yards.

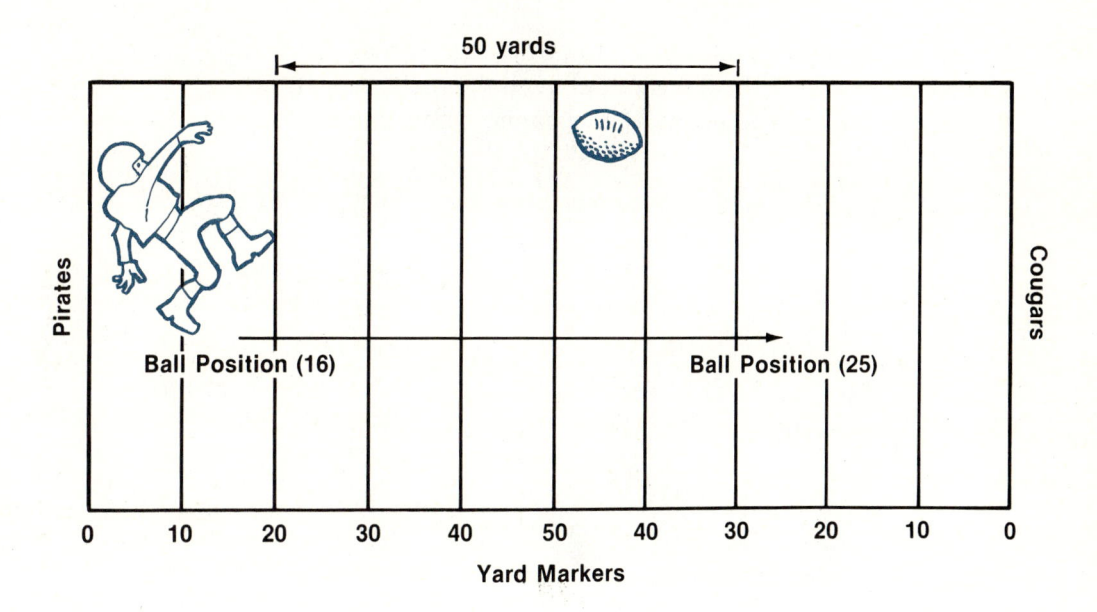

Draw pictures or diagrams to help you solve the following problems.

1. Ed is counting his calories. He came home and found $\frac{1}{2}$ of a pizza. He ate $\frac{1}{3}$ of what he found. A whole pizza is 1,800 calories.

    A. What fraction of the whole pizza did he eat? _____

    B. How many calories did he eat? _____

UNIT 2 ■ WORD PROBLEMS AND PROBLEM SOLVING

2. Liz and Randy are making picture frames. Each frame measures $4\frac{1}{2}$ inches by $10\frac{1}{2}$ inches. If they plan to make 12 frames, how much framing material will they need to buy? _____

3. Six 2-inch by 3-inch pictures are to be arranged on an 18-inch by 15-inch page so that the distance between the pictures is the same as the distance to the edges of the paper. Show the layout.

4. Jan and Jessie are going to wallpaper 3 walls in their living room. How many square feet of wallpaper will they need if one wall measures 8 feet by 12 feet, the second wall measures 8 feet by 10 feet (and has a door that measures 3 feet by 7 feet), and the third wall measures 8 feet by 10 feet (and has a window that measures 5 feet by 3 feet). _____

5. The local deli is ordering cheese and can purchase a small block of cheddar cheese for $12.50. If the price per pound is the same, what will the deli pay for a block of cheese that is twice as tall, twice as wide, and twice as long as the small block? _____

STRATEGY 9—MANIPULATE (TOUCH AND MOVE) OBJECTS

Sometimes it is helpful to *manipulate*, or touch and move, things with your hands to solve a problem. In Example 1, removing the toothpicks makes the problem easier to solve.

Example 1   Arrange 24 toothpicks to make the following shape:

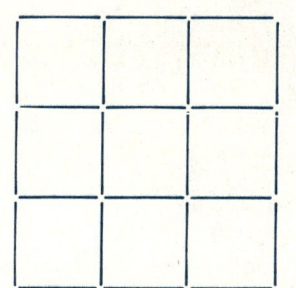

Remove 4 toothpicks and leave 5 squares.

Solution

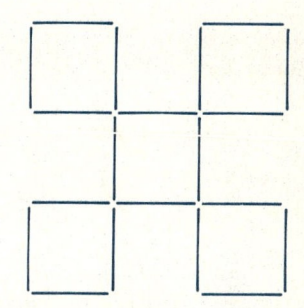

Example 2   Trace, cut out, then manipulate these shapes to form the letter *L*.

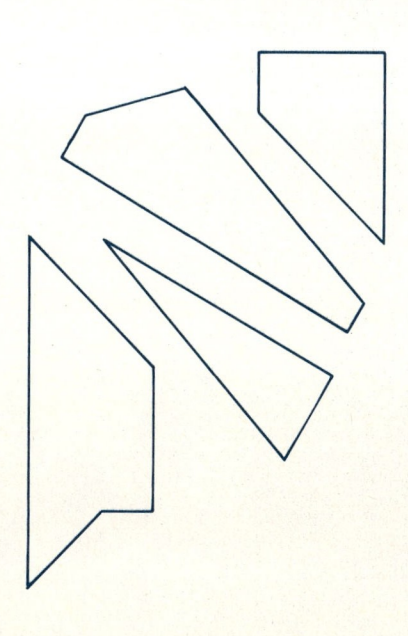

Solution

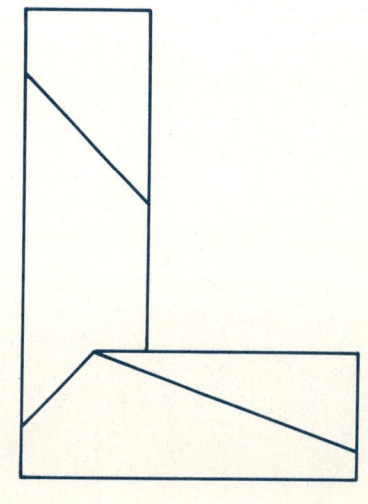

UNIT 2 ■ WORD PROBLEMS AND PROBLEM SOLVING

Solve the following problems. Draw your answers in the spaces provided.

1. Arrange 24 toothpicks to make the shape at the right.

   A. Remove 4 toothpicks and leave 7 squares.
   B. Remove 8 toothpicks and leave 5 squares.

2. Arrange 16 toothpicks to make the shape at the right.

   A. Add 5 toothpicks to make 10 triangles.
   B. Add 4 toothpicks to make 6 squares.

3. Arrange 18 toothpicks to make the shape at the right.

   A. Take 4 toothpicks away and leave 3 squares.
   B. Take 6 toothpicks away and leave 3 squares.

   C. Take 7 toothpicks away and leave 3 squares.

4. Arrange 8 toothpicks to make the shape at the right.

   A. Add 4 toothpicks to make 6 triangles.

   B. Add 4 toothpicks to make 4 diamonds.

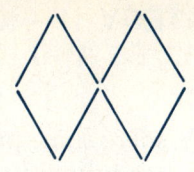

5. Arrange 16 toothpicks to make the shape at the right.

   A. Move 4 toothpicks to make 5 squares.

   B. Add 5 toothpicks to make 10 triangles.

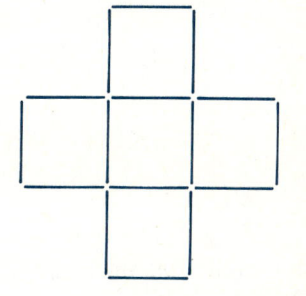

6. Trace, cut out, and then manipulate these shapes to form a capital *E*. Draw your arrangement in the space at the right.

UNIT 2 ■ WORD PROBLEMS AND PROBLEM SOLVING

STRATEGY 10—KEEP TRACK OF CLUES AND INFORMATION

To solve some problems, it is helpful to keep track of clues and to organize information. Keeping a tally, making a grid, or drawing a Venn diagram can help to organize information. Sometimes you may need to read a problem several times to find all the clues and keep track of the information.

Solve the following problems by organizing the information given.

1. Mark, Melinda, and Marvin play the trumpet. Marla and Mitchell play the trombone, and Oliver plays the tuba. The music teacher said that in order for the orchestra to sound its best, they need 3 times as many woodwinds as brass instruments and 3 times as many strings as percussions. Sharlett, Sherry, and Steve play the drums.

   A. How many woodwinds are needed? _____

   B. How many strings are needed? _____

   C. How many will be in the orchestra? _____

2. Complete the family tree by organizing the given information.

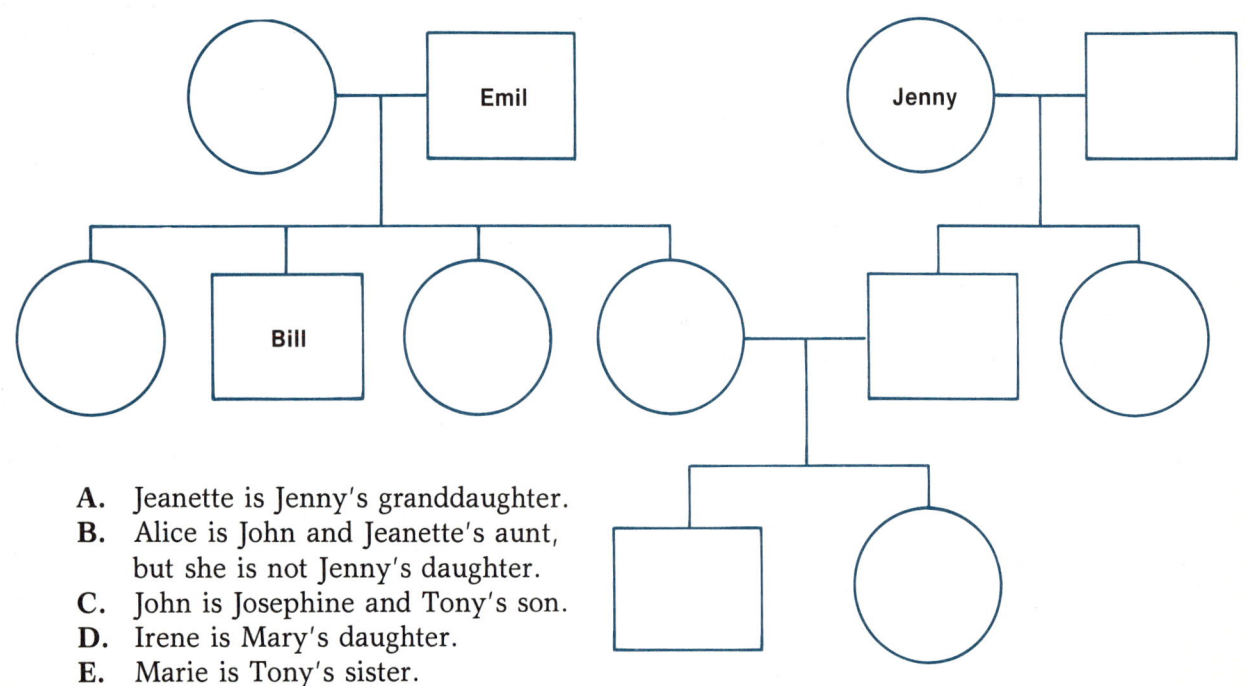

   A. Jeanette is Jenny's granddaughter.
   B. Alice is John and Jeanette's aunt, but she is not Jenny's daughter.
   C. John is Josephine and Tony's son.
   D. Irene is Mary's daughter.
   E. Marie is Tony's sister.
   F. John Senior is Jeanette's grandfather.

92   UNIT 2 ■ WORD PROBLEMS AND PROBLEM SOLVING

Tally

To solve some problems, it is helpful to *tally*, or record and count, the given information.

In the following problem, 11 friends voted on the games they liked best. They had 3 choices, baseball, kickball, and soccer, and could choose 2 of the 3. The tally has been started. Complete the tally in order to answer the questions that follow.

Ken	**Joanne**	**Jeff**	**Brad**	**Josie**	**Noreen**
baseball	kickball	soccer	soccer	baseball	soccer
kickball	baseball	baseball	kickball	kickball	kickball

Maggie	**Keith**	**Joe**	**Jeanine**	**Beth**
baseball	soccer	baseball	baseball	baseball
soccer	kickball	kickball	soccer	soccer

	Baseball	Kickball	Soccer
Boys	/	/	
Girls	/	/	

1. Which game received the most votes from the girls? _____
2. Which game received the most votes from the boys? _____
3. Which game received the most votes from the 11 friends? _____

Grid

A *grid* is made up of horizontal and vertical lines and can be used to keep track of information.

Example   Gwen, Duane, and Harry each ordered something different at a snack bar. One had a hot dog, one had a hamburger, and one had a milk shake.

A. Gwen did not have a hot dog.
B. Harry did not get a plate.

What did each person have?

	Hot Dog	Hamburger	Milk Shake
Gwen	No	Yes	No
Duane	Yes	No	No
Harry	No	No	Yes

Gwen ___hamburger___
Duane ___hot dog___
Harry ___milk shake___

UNIT 2 ■ WORD PROBLEMS AND PROBLEM SOLVING

Solve the following problems by organizing your information in a grid.

1. Five friends were playing basketball. One had a knee brace, one wore a headband, one wore yellow shorts, one put his shirt on the bench, and one doesn't like coed sports. Use the clues below and a grid to match the names and descriptions.

   A. Peter hates yellow and has had no injuries this year.
   B. Sonia has long hair that gets in her eyes.
   C. Angela prefers to play basketball with the girls.
   D. Leon has been unable to play basketball for the last two weeks.
   E. Ben and Angela are dating.

   Peter   _____
   Sonia   _____
   Angela   _____
   Leon   _____
   Ben   _____

2. Henry, Jill, Rob, and Doris each have a collection. They collect matchbooks, unicorns, cassette tapes, and concert posters. Henry has the walls of his room decorated. Jill likes anything to do with fantasy and science fiction. Doris' grandparents add to her collection whenever they travel. What does Rob collect? _____

3. Mei-Ling, Lee, and Sylvia were planning a party. Each chose *two* of their favorite foods to serve. Use the clues below and a grid to decide which of the following each person chose.

   chips and guacamole   salami   Mei-Ling _____
   carrot sticks   bread   Lee _____
   ice cream   cheese   Sylvia _____

   A. Mei-Ling loves Mexican food.
   B. Sylvia would be happy to eat sandwiches three times a day and is allergic to milk.
   C. Lee is on a diet, but can't resist ice cream.

4. Miriam, Edie, and Nancy are each going to the dance with guys who play different sports. The girls love to watch their dates play their sports. The sports are baseball, basketball, and football. The athletes are Bob, Vince, and Louis. *Use one or more grids* to keep track of the clues and discover who will be going to the dance together.

   A. Miriam loves spring weather.
   B. Nancy has allergies and likes to stay indoors.
   C. Louis had to miss his vacation for training.
   D. Bob must wear court shoes (high tops).

UNIT 2 ■ WORD PROBLEMS AND PROBLEM SOLVING

Venn Diagram

A Venn diagram is a series of intersecting circles or squares used to gather, sort, or classify information.

Example   The Fosters are planning a family reunion and want to organize information about the participants. Jake is single, will be traveling 250 miles, and has no children. Melanie will be traveling 502 miles, is married, and has no children. Bud will be traveling 8 miles, has children, and is married. Ginger is married, will be traveling 8 miles, and has children. Rich will be traveling 38 miles, is divorced, and has children. Judy is married, has children, and will be traveling 158 miles. Sally has a child, will be traveling 32 miles, and is single. Patsy is single, will be traveling 9 miles, and has children.

This information has been recorded in the Venn diagram as shown. Note that each circle category is labeled, and each name is placed in the appropriate circle. Some names must be placed where the circles overlap because they fit into more than one category.

The Venn diagram can be used to answer questions about the participants at the family reunion.

How many people are married? __4__

How many people are single? __4__

Who can be found in all 3 circles? __Judy__

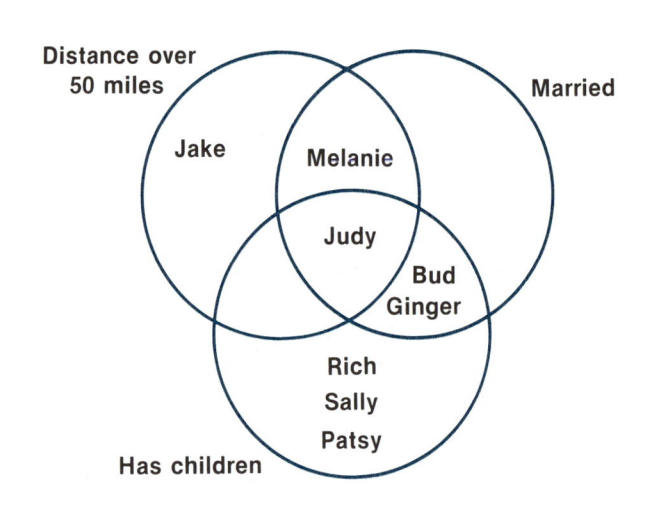

Solve the following problems using Venn diagrams.

1. Sort the numbers 1 through 25 in the appropriate squares of the Venn diagram below. If a number does not belong in any square, write it outside the Venn diagram. Numbers 1 through 4 have already been sorted for you.

    A. How many prime numbers are shown?

    _____

    B. How many odd numbers are shown that are also multiples of 5?

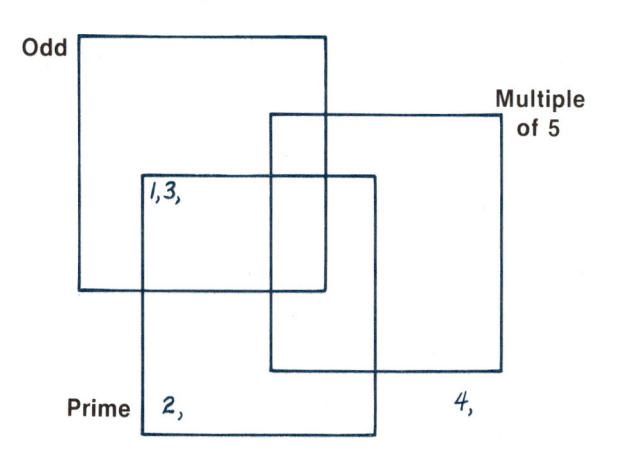

    _____

96   UNIT 2 ■ WORD PROBLEMS AND PROBLEM SOLVING

2. Ten students parked their cars off campus because the cars would not pass the safety inspection. Sort the cars in the Venn diagram and answer the questions below.

   George—no seat belt, broken horn
   Karen—broken muffler
   Eric—broken horn
   Terry—no seat belt, broken horn
   Debbie—broken horn, broken muffler
   Justin—no seat belt
   Lily—broken muffler, no seat belt
   Bruce—broken muffler, broken horn, no seat belt
   Rhonda—no seat belt, broken muffler
   Joel—no seat belt, broken horn

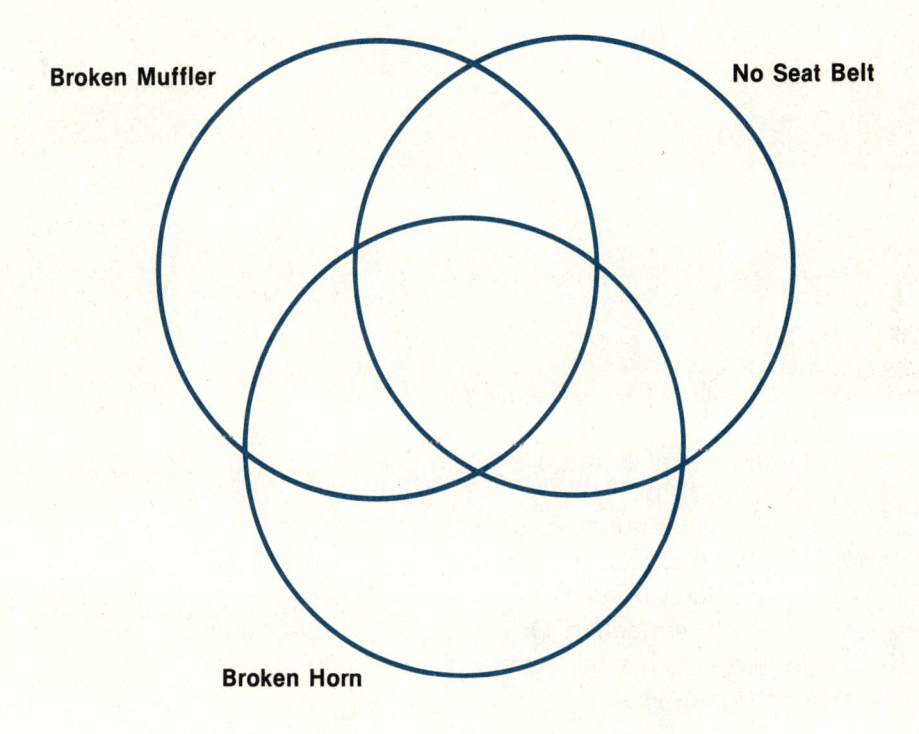

   A. How many different kinds of car problems and combinations of car problems are illustrated in the Venn diagram? _____

   B. List the categories and note how many cars fit in each category.

UNIT 2 ■ WORD PROBLEMS AND PROBLEM SOLVING

3. Make a Venn diagram to sort the numbers 1 through 30 according to whether they are even numbers, multiples of 3, or multiples of 5. If a number does not belong to one of these categories, write it outside the Venn diagram.

   A. How many multiples of 3 are even numbers? _____

   B. How many multiples of 5 are even numbers? _____

   C. How many multiples of 5 and 3 are odd numbers? _____

4. Margarite and Rodney were taking an inventory of a department store's shirts. There were 25 shirts. Nineteen shirts had a single pocket, 11 shirts had button-down collars, 2 shirts had neither a pocket nor a button-down collar. Of the 25 shirts, how many had both a single pocket and a button-down collar? Use a Venn diagram to answer this question. _____

{3}
COMPARISON SHOPPING

SKILLS YOU WILL NEED

    Basic math skills

SKILLS YOU WILL LEARN

    Finding unit prices
    Shopping for convenience products
    Finding cost per serving
    Making miscellaneous purchases
    Figuring automobile expenses
    Figuring the cost of credit

SITUATIONS IN WHICH YOU'LL USE THESE SKILLS

    Shopping for groceries and other purchases
    Buying an automobile
    Using credit

Comparison shopping is a skill. You will be more satisfied with your purchases if you develop and use this skill. Whether you're buying groceries, clothes, or a car, comparison shopping can help you make a wise decision. Before making any purchase, you might ask yourself the following questions:

1. Why do I want to make this purchase?
2. What quality do I need?
3. Where shall I buy it?
4. Should I compare prices at other stores?
5. What does the ad really mean?
6. What is the real cost?

Each of these questions is important, but the final decisions for your personal buying are up to you. This unit will present math skills to assist you in making purchase and finance decisions.

SHOPPING FOR GROCERIES

Comparison shopping helps you get the most value for your dollar. Shopping for groceries includes determining your needs, finding unit prices, considering convenience products, and finding cost per serving or use.

Your answers to the following questions can help you determine your food and household needs.

1. How much money is available?
2. What size is your family?
3. Who will be cooking?
4. How much time is available for preparation?
5. How long will the food keep?
6. How much storage space is available?
7. What quality is required?

Each of these questions is important. Perhaps the most difficult one to answer is the last one. Determining the quality you need can be a challenge. You may have to decide whether to buy a canned, fresh, frozen, or convenience product, or whether to buy an expensive or inexpensive cut of meat. Determining quality also involves taste, appearance, and use. For example, using a can of peach halves for a pie doesn't make sense. Peach halves are more expensive than peach slices, and appearance is less important because they will be covered by a crust. In this case, the taste and cost of the peaches determines the quality you need. In the same way, a stew does not require buying an expensive steak.

Unit Pricing

Comparing costs of products is sometimes difficult. Which is the better buy? The smaller box or the bigger box?

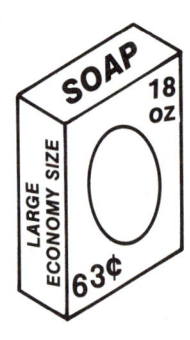

To find the better buy, you need to find the unit price. The *unit price* is the cost per market unit. Common market units are ounces, pounds, pints, quarts, grams, and liters. Some stores display unit prices next to items; other stores do not. It is important to learn how to find unit prices so you can compare various sized items.

To find the unit price of an item, divide the total price by the number of units.

$$\frac{\text{unit price}}{\text{number of units} \overline{)\text{total price}}}$$

Example 1 Which box of soap on page 100 is the better buy?

The 13-ounce box of soap is 39¢, so the unit price is 3¢.

The 18-ounce box of soap is 72¢, so the unit price is 4¢.

$$\frac{3¢}{13\overline{)39¢}} \text{ cost per ounce}$$

$$\frac{4¢}{18\overline{)72¢}} \text{ cost per ounce}$$

In this case, the large box of soap costs more per ounce, so the small box is the better buy.

Answer the following questions.

1. A small box of raisins holds 3 ounces and sells for 57¢. A large box of raisins holds 5 ounces and sells for 80¢.

   A. What is the cost per ounce of the small box? _____

   B. What is the cost per ounce of the large box? _____

   C. Which is the better buy? _____

2. Find the unit prices for these cans of tuna:
   A. 9-ounce can for 72¢ _____

   B. 6-ounce can for 54¢ _____

UNIT 3 ■ COMPARISON SHOPPING

Stores often sell several cans of an item for one price, such as 2 cans for $1.46. In this case, the unit price is the price of a single can. To find the unit price, divide the total price by the number of cans.

Example 2   What is the unit price for peaches marked 2 cans for $1.46?

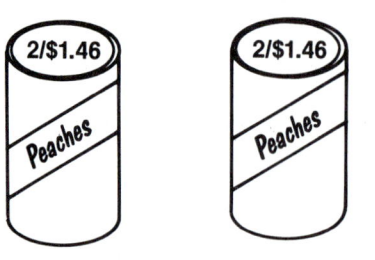

$$\begin{array}{r}\$0.73 \\ 2\overline{)\$1.46}\end{array} \text{ per can}$$

Sometimes an item will be priced in two different ways. Usually the cost of the single item is more than the cost of the same item in the combination price.

Example 3   How much is saved per can?

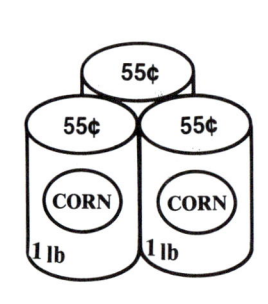

Cost per can if 1 is purchased:
55¢

Cost per can if 3 are purchased:

$$\begin{array}{r}\$0.50 \\ 3\overline{)\$1.50}\end{array} \text{ per can}$$

Answer: 55¢ − 50¢ = 5¢ saved per can

When stores use combination prices, a single item may cost a fraction of a cent. For example, cans of fruit may be priced at 3 for $1.69. If you only buy 1 can, the store will charge you 57¢ even though $1.69 ÷ 3 = 56\frac{1}{3}$¢. The store always rounds to the next *larger* cent.

UNIT 3 ■ COMPARISON SHOPPING

Answer the following questions. Round your answers to the next *larger* cent, using the store method of rounding.

1. String beans are advertised at 2 cans for $1.25. What is the price of 1 can? _____

2. A store is selling corned beef at 5 cans for $5.50 or 1 can for $1.19.

   A. What is the unit price of 1 can when purchasing 5 for $5.50? _____

   B. How much is saved per can when buying 5 cans? _____

3. If 2 cans of fruit cocktail sell for $1.26, how much will 1 can cost? _____

4. If automobile oil costs $3.50 for 3 cans, how much will 10 cans of oil cost? _____

5. At 2 for 58¢, find the cost of 8 candy bars. _____

UNIT 3 ■ COMPARISON SHOPPING

Find the unit prices in the following problems. Use the store method of rounding to the next larger cent.

1. A 4-pound beef roast costs $7.98. A 3-pound chicken costs $2.75.

   A. How much is the beef per pound? _____

   B. How much is the chicken per pound? _____

2. Oranges cost $1.40 for a 3-pound sack at Savemore, 45¢/pound at Farmland, and $8.59 for a 20-pound case at Kramers. Find the unit prices per pound.

   A. 3 pounds for $1.40 _____

   B. 45¢/pound _____

   C. 20-pound case at $8.59 _____

3. Chocolate chips cost $1.59 for 12 ounces. Chocolate-flavored pieces cost 65¢ for 7 ounces. What is the unit price for each?

   A. Chocolate chips _____

   B. Chocolate-flavored pieces _____

4. What is the unit price if $4\frac{1}{2}$ pounds of tomatoes cost $1.26? _____

5. If $3\frac{1}{4}$ pounds of cherries cost $2.89, what does 1 pound cost? _____

6. Cereal sells for $1.05 for a 12-ounce box. What is its unit price? _____

7. Ten pounds of potatoes cost 85¢ and you get 31 potatoes in a bag.
   A. How much are potatoes per pound? _____

   B. How much does 1 potato cost? _____

8. If a store is selling corn for 7 ears for $1, how much would 1 ear of corn cost? _____

UNIT 3 ■ COMPARISON SHOPPING

Items are often packaged in dissimilar units. How can you compare 1 quart and 24 ounces?

To compare quarts and ounces, you must change the quarts to equivalent ounces. The table below shows the equivalent units you need to find unit prices.

> 1 gallon = 128 fluid ounces = 4 quarts
> 1 quart = 32 fluid ounces = 2 pints
> 1 pint = 16 fluid ounces = 2 cups
> 1 pound = 16 ounces

To change 1 pint 2 ounces to ounces, convert the 1 pint to 16 ounces and add the 2 ounces to get 18 ounces.

Which amount is larger? If the two amounts are equal, write "equal" in the blank.

1. 18 ounces or 1 pound _____

2. 12 ounces or $\frac{1}{2}$ pound _____

3. 2 pints or 1 quart _____

4. 3 ounces or $\frac{1}{4}$ pound _____

5. $\frac{1}{2}$ pound or 6 ounces _____

Use your calculator to find the unit prices in the following problems. Use the store method of rounding to the next *larger* cent.

1. One quart of mayonnaise costs $1.59, and a 24-ounce jar costs $1.49.

   A. What is the unit price of the quart? _____

   B. What is the unit price of the 24-ounce jar? _____

2. Find the unit prices for two sizes of liquid floor cleaner.

   A. 32 fluid ounces for 89¢ _____

   B. 1 pint 2 ounces for 79¢ _____

106 UNIT 3 ■ COMPARISON SHOPPING

3. Find the unit prices for three sizes of floor wax.
   A. 36 fluid ounces for $1.85   _____

   B. 1 pint 11 ounces for $1.09   _____

   C. 1 quart 14 ounces for $1.39   _____

4. Find the unit prices for two sizes of catsup.
   A. 1 quart 10 ounces for 99¢   _____

   B. 28 fluid ounces for 89¢   _____

If you wish to find the unit price of something weighing 1 pound 2 ounces, first convert the 1 pound to 16 ounces. Then add the 2 ounces to get 18 ounces.

Estimate the unit price for the following problems. Check your answer with the calculator. Use the store method of rounding to the next larger cent.

1. Find the unit prices for:
   A. A package of rice weighing 1 pound 5 ounces that costs 65¢.

                                    Estimated   _____
                                    Calculated   _____

   B. A package of rice weighing 5 ounces that costs 27¢.

                                    Estimated   _____
                                    Calculated   _____

2. Find the unit prices for:
   A. A package of macaroni weighing 2 pounds 10 ounces that costs $1.59.

                                    Estimated   _____
                                    Calculated   _____

   B. A package of macaroni weighing 14 ounces that costs 89¢.

                                    Estimated   _____
                                    Calculated   _____

UNIT 3 ■ COMPARISON SHOPPING

3. Find the unit prices for:

   A. A $1\frac{1}{2}$-pound package of popcorn at $1.55.

   Estimated _____

   Calculated _____

   B. A 14-ounce package of popcorn at 75¢.

   Estimated _____

   Calculated _____

Sometimes you can save money while shopping by using coupons to purchase needed items. The wise consumer will compare the price using the coupon to the price of other brands to verify the savings.

Use your calculator to solve the following problems. Use the store method of rounding to the next larger cent.

1. Find the unit prices for:

   A. A 3-ounce package of walnuts that regularly sells for $1.18 if a 15¢ coupon is used. _____

   B. A 10-ounce package of walnuts at $2.39. _____

   C. Bulk walnuts at $3.39 a pound. _____

2. Find the unit prices for:

   A. A package of oatmeal weighing 1 pound 2 ounces that regularly costs $1.35 if a 10¢ coupon is used. _____

   B. A package of oatmeal weighing 2 pounds 10 ounces that costs $2.53. _____

3. Find the unit prices for:

   A. A box of laundry soap weighing 17 ounces that costs $1.15. _____

   B. A box of laundry soap weighing 2 pounds 10 ounces that regularly costs $2.60 if you use a 20¢ coupon. _____

   C. A box of laundry soap weighing 4 pounds 8 ounces that costs $4.15. _____

Convenience Products

Convenience products, such as frozen dinners, are designed to save time and effort. You usually must pay for this convenience, so convenience products are often more expensive. You must decide whether the time and effort saved is worth the extra expense.

When necessary, use the store method of rounding money to answer the following questions.

1. A 3-pound whole chicken costs $2.64. A $3\frac{1}{2}$-pound cut-up chicken costs $3.45. How much per pound do you save by cutting up the chicken yourself? _____

2. One dozen cloth diapers cost $12.50, and can be used for 2 years.
   A. If the diapers are changed 8 times a day for 2 years, how many diaper changes are made? _____

   B. How much will each change of a diaper cost? _____

UNIT 3 ■ COMPARISON SHOPPING

3. Disposable diapers are more convenient than cloth diapers. Forty-eight disposable diapers sell for $9.34.

   A. How much do disposable diapers cost per diaper? _____

   B. Using your answer to Problem 2B, how much per use can be saved by using cloth diapers instead of disposable diapers? _____

   C. If the baby is changed 8 times a day for 2 years, how much will be spent for disposable diapers in 2 years? _____

4. One dozen brown and serve biscuits sell for 96¢. Ten refrigerator biscuits sell for 99¢. Using a biscuit mix, 1 dozen biscuits cost 44¢ to prepare. Find the unit price for each.

   A. brown and serve _____

   B. refrigerator _____

   C. biscuit mix _____

Cost Per Serving

In order to compare costs, it is sometimes necessary to know the *cost per serving*. The cost per serving is calculated the same way as the unit price: divide the number of servings into the total price.

$$\text{number of servings} \overline{\smash{\big)}\ \text{total price}}^{\text{cost per serving}}$$

Use your calculator to solve the following problems. Use the store method of rounding to the next larger cent.

1. Find the cost per serving for the following sizes of milk, if 1 serving is 1 cup (8 ounces).

    A. 1 quart at 54¢ _____

    B. 1 half-gallon at 96¢ _____

    C. 1 gallon at $1.87 _____

2. Find the cost per $\frac{1}{2}$-cup serving for the following orange juice choices.

    A. Canned juice at $1.49 for 1 quart 12 ounces _____

    B. Frozen juice that makes $1\frac{1}{2}$ quarts and costs $1.15 _____

    C. Freshly squeezed orange juice if 3 oranges cost 72¢ and 1 orange makes $\frac{1}{2}$ cup _____

UNIT 3 ■ COMPARISON SHOPPING

When food is prepared from a recipe, the cost per serving depends on the cost of the ingredients. To find the cost per serving for a recipe, find the total cost of the ingredients and divide by the number of servings.

For the following problems, use the costs in the table to find the cost per recipe and cost per serving. Use your calculator and the store method of rounding.

eggs (1)	$0.08	pepperoni (8 ounces)	$1.99
flour (1 cup)	0.08	shortening ($\frac{1}{2}$ cup)	0.16
margarine (1 cup)	0.30	sugar (1 cup)	0.16
milk (1 cup)	0.14	tomato sauce (8 ounces)	0.22
mozzarella cheese (6 ounces)	1.49	yeast (1 package)	0.25
mushrooms (8 ounces)	1.00	unsweetened chocolate (1 square)	0.28
oil (1 cup)	0.32		

baking powder, baking soda, salt, spices, flavoring (*each use*)   0.04

1. Brownies: Makes 12 servings

    $\frac{1}{2}$ cup margarine
    1 cup sugar
    1 teaspoon vanilla
    2 eggs
    2 squares chocolate
    $\frac{3}{4}$ cup flour

    Total cost _____

    Cost per serving _____

2. Kevin is planning a pizza dinner and has decided on a mozzarella, mushroom, and pepperoni combination. He has three choices for preparing the pizza and decides to compare prices for his three possibilities. Find the total cost for each of the following.

    A. Package mix, not including cheese, meat, or mushrooms is $1.30. Include prices for 6 ounces of cheese, 4 ounces of pepperoni, and 4 ounces of mushrooms to find the total cost.

    _____

112   UNIT 3 ■ COMPARISON SHOPPING

B. Kevin's sister, Heather, uses the following pizza recipe. Find the total cost using this recipe.

_____

```
                       PIZZA
        Crust                    Topping

   1 package yeast        8-ounce can tomato sauce
   3/4 cup water          1/2 teaspoon oregano
   3 1/3 cups flour       1/2 teaspoon rosemary
                          1/2 teaspoon salt
                          1/4 teaspoon pepper
                          4 ounces mushrooms
                          4 ounces pepperoni
                          6 ounces mozzarella cheese

   Mix crust.  Cover and let rise in warm place until double, about 1
   hour.  Pat dough into bottom of greased pizza pan.  Mix tomato sauce
   and seasonings.  Spread on dough.  Top with mushrooms, mozzarella,
   and pepperoni.  Bake at 375° for 20 minutes.
```

C. How much more would Kevin spend if he decided not to cook Heather's recipe and sent out for a pepperoni, mushroom, and cheese pizza that costs $8.95?

_____

MISCELLANEOUS PURCHASES

Every time you buy something you must make decisions. You want to get a product you will like and use. You also don't want to spend more money than necessary. Sometimes it is difficult to determine the wisest purchase. Advertisements can help, but may also be confusing or misleading. Math skills can help you understand advertisements so you can make the wisest purchases.

UNIT 3 ■ COMPARISON SHOPPING

Solve the following problems to determine the wisest purchases.

1. Sure-Fit Factory is advertising a sale on jeans. When one pair of jeans is purchased for $24.90, the second pair of jeans is half price. Jeans Outlet is advertising the same jeans for $26 reduced by 20%.

   Sure-Fit Factory

   Buy 1 Pair At
   $24.90
   Get 2nd Pair At
   HALF PRICE!!

   JEANS OUTLET

   SAVE 20%
   On Our $26 Jeans

   !! SALE !!

   A. What will two pairs of jeans cost at Sure-Fit Factory? _____

   B. What will two pairs of jeans cost at Jeans Outlet? _____

2. Jordan has been buying his favorite magazine each month at the store for $2.25. His friend Scott subscribes to the same magazine. Scott's $1\frac{1}{2}$-year subscription costs $38.70.

   A. How much does Scott pay per month for the magazine? _____

   B. How much does Scott save in $1\frac{1}{2}$ years by subscribing to the magazine? _____

114   UNIT 3 ■ COMPARISON SHOPPING

3. Style-for-Less is advertising a 65% savings on winter coats. The Clothes Horse is advertising winter coats for $\frac{1}{2}$ the regular price, plus an additional 20% off that sale price. What is the sale cost of a $139 coat at each store?

   A. Style-for-Less _____

   B. Clothes Horse _____

The Clothes Horse

Take HALF OFF the regular price, then take an ADDITIONAL 20% OFF

50% OFF
+ 20% OFF
───────
HUGE SAVINGS

COAT SALE

4. Christy plans to buy a new portable transistor radio. Save-All advertises a sale of $\frac{1}{3}$ off the regular $72.99 price. Super Saver offers a coupon that will save $15 on their $68.99 model. What is the sale price at each store?

   A. Save-All _____

   B. Super Saver _____

UNIT 3 ■ COMPARISON SHOPPING

AUTOMOBILE EXPENSES

Buying a car is a major consumer decision. Automobile expenses include more than just the initial cost of the car. Car insurance, gasoline expense, and maintenance costs are also important considerations.

Solve the following problems. Use the car insurance table below when necessary.

CAR INSURANCE COSTS		
	Rate Per Quarter Year	*Deductable Amount
Chevrolet Monte Carlo, 1976	$200.00	$250
Honda Civic, 1976	$162.50	$200
MG Midget, 1975	$262.50	$250
Triumph Spitfire, 1971	$300.00	$200

*The amount the owner pays before an insurance company pays the balance of an accident damage claim.

1. Compare the annual costs of ownership for the following automobiles. Assume each is driven 75 miles per week and gasoline costs 96¢ a gallon. Also assume that maintenance will be the same for the vehicles and will cost $250 per year. Since this is an estimate, round to the nearest dollar.

> HONDA CIVIC '76
> 4 speed; one owner; $650
> new tires; 35 mpg

Initial cost _____
Annual gasoline cost _____
Annual insurance cost _____
Annual maintenance _____
Total first year cost _____

> CHEVY MONTE CARLO '76
> Runs great; 14 mpg; $795
> Call after 6 pm

Initial cost _____
Annual gasoline cost _____
Annual insurance cost _____
Annual maintenance _____
Total first year cost _____

> TRIUMPH '71 Spitfire
> Convertible; body needs
> work $700; 28 mpg

Initial cost _____
Annual gasoline cost _____
Annual insurance cost _____
Annual maintenance _____
Total first year cost _____

> MG '75, MIDGET
> Must sell; runs good
> $750; 18 mpg; ask for Bud

Initial cost _____
Annual gasoline cost _____
Annual insurance cost _____
Annual maintenance _____
Total first year cost _____

2. Choose one of the four cars. How much should be budgeted per month for gas, maintenance, and insurance? (Round to the nearest dollar.)

   A. Car of your choice _____

   B. Monthly cost to be budgeted _____

3. Frequently, insurance companies will offer a discount if the insurance premium is paid once a year. If the rate is reduced 10%, what is the annual cost of insurance for each car?

   A. Chevy _____

   B. Honda _____

   C. MG _____

   D. Triumph _____

4. The 1976 Monte Carlo was in an accident and the repair bill was $435. What amount did the insurance company pay the owner? _____

UNIT 3 ■ COMPARISON SHOPPING

COST OF CREDIT

Credit is a convenience, and, as such, it costs money. It's important to shop for credit because the cost of credit can vary greatly. Understanding how credit costs are figured can save money.

Solve the following problems.

1. Compare the finance charges for a $1,000 car loan from a bank, a credit union, and a finance company. To find the amount of interest, multiply the annual percentage interest rate (APR) times the amount borrowed.

	APR	Dollar Amount of Interest Charged	Total Repayment
Credit Union	18%		
City Bank	15%	$150	$1,150
Finance Company	21%		

2. Compare the finance charge of 21% APR for a $1,000 loan from the credit union for the following lengths of time. Multiply the interest amount for 1 year times the length of the loan (in years) to find the total amount of interest.

Credit Union		
Length of Loan	Dollar Amount of Interest Charged	Total Repaymant to Credit Union
1 year		
18 months		
2 years		
3 years		

3. Lillian has $1,650 in her savings account for college expenses. She earns 5.5% interest on that amount in 1 year. She wants to buy a car for $1,650. Her parents will not let her use her college money, but will cosign a bank loan for 1 year at 17%.

   A. How much money will she earn from her savings account in 1 year? _____

   B. How much interest will she pay for the bank loan? _____

   C. What is the difference between the cost of borrowing $1,650 for 1 year and the amount earned by saving $1,650 for 1 year? _____

4. Daisy and Rick want to borrow $1,400 to furnish their apartment. The furniture store will finance the $1,400 at 19% interest for 2 years. Their credit union will charge 15% for 3 years.

   A. What is the finance charge at the furniture store? _____

   B. What is the finance charge at the credit union? _____

5. One generally accepted recommendation is that you limit the total amount borrowed to less than 20% of your take-home pay. If your take-home pay is $1,050 per month, what is the most you should borrow in 1 year? _____

UNIT 3 ■ COMPARISON SHOPPING

When you use a credit card at a department store, the store will charge you interest each month on any unpaid balance. Department stores have different ways of calculating the amount of interest owed on an unpaid balance.

1. *Previous balance* The finance charge is calculated on what was owed last month, before subtracting any payment that might have been made during the billing period.
2. *Average daily balance* The finance charge is calculated on the average between what was owed last month and what is now owed.
3. *Adjusted balance* The finance charge is calculated on the balance after payments made during the billing period have been subtracted.

	Previous Balance	Average Daily Balance	Adjusted Balance
Monthly Interest Rate	1.8%	1.8%	1.8%
Previous Balance	$600	$600	$600
Payments	$250	$250	$250
Interest Charge	$10.80 ($600 × 1.8%)	$8.55 ($475 × 1.8%)*	$6.30 ($350 × 1.8%)

*$600 + ($600 − $250) = $950, $950 ÷ 2 = $475

Kelly owed $735 on a department store credit card in February. Kelly paid $50 on that bill. The March billing shows a balance of $685. The monthly interest rate is 1.8%. Study the table and definitions above to help you answer the following questions.

1. If the *previous balance* method is used, on what amount will the finance charge be figured? _____

2. What will the finance charge be for the amount in Problem 1? _____

120 UNIT 3 ■ COMPARISON SHOPPING

3. If the *average daily balance* method is used, on what amount will the finance charge be figured? _____

4. What will the finance charge be for the amount in Problem 3? _____

5. If the *adjusted balance* method is used, on what amount will the finance charge be figured? _____

6. What will the finance charge be for the amount in Problem 5? _____

4
TABLES, GRAPHS, AND AVERAGES

SKILLS YOU WILL NEED

    Basic math skills
    Willingness to analyze visual material

SKILLS YOU WILL LEARN

    Reading tables
    Reading graphs
    Making graphs from given information
    Finding averages: mean, median, and mode

SITUATIONS IN WHICH YOU'LL USE THESE SKILLS

    Reading newspapers and magazine articles
    Studying for other classes
    Reading transportation schedules
    Reading tax charts

Today we live in an information age. The technology for gathering, displaying, and using information has advanced greatly. The result is an increased use of tables, graphs, and averages in the media and in our daily lives. This unit will give you practice in understanding, using, and interpreting information.

READING TABLES

A table lists information in rows and columns. When working with tables, always read the title and labels for each column.

Example How many calories are in 1 ear of corn?

Calories of Common Foods		
Food	Quantity	Calories
chili	1 small bowl	250
beans, green	⅔ cup	18
doughnut	1	250
corn	1 ear	92
hot dog	1	125
hot dog bun	1	125
potato	1	83
ham	3 ounces	179
steak	3.36 ounces	212
orange	1	65

→ 1 ear of corn contains 92 calories

HINT: It is sometimes helpful to hold a paper or ruler under the line to make sure you are reading straight across.

Use the tables to answer the following questions.

1. Using the tax table below, find the tax for these amounts.

   A. $17.86 _____

   B. $8.92 _____

   C. $0.77 _____

   D. $10.20 _____

   E. What is the largest transaction amount in this table?
   _____

   F. What is the tax on the amount in part E?
   _____

   G. What is the greatest amount you can spend without paying tax?
   _____

6% SALES TAX REIMBURSEMENT SCHEDULE					
Transaction	Tax	Transaction	Tax	Transaction	Tax
0.01- 0.10	0.00	8.42- 8.58	0.51	16.92-17.08	1.02
0.11- 0.22	0.01	8.59- 8.74	0.52	17.09-17.24	1.03
0.23- 0.39	0.02	8.75- 8.91	0.53	17.25-17.41	1.04
0.40- 0.56	0.03	8.92- 9.08	0.54	17.42-17.58	1.05
0.57- 0.73	0.04	9.09- 9.24	0.55	17.59-17.74	1.06
0.74- 0.90	0.05	9.25- 9.41	0.56	17.75-17.91	1.07
0.91- 1.08	0.06	9.42- 9.58	0.57	17.92-18.08	1.08
1.09- 1.24	0.07	9.59- 9.74	0.58	18.09-18.24	1.09
1.25- 1.41	0.08	9.75- 9.91	0.59	18.25-18.41	1.10
1.42- 1.58	0.09	9.92-10.08	0.60	18.42-18.58	1.11
1.59- 1.74	0.10	10.09-10.24	0.61	18.59-18.74	1.12
1.75- 1.91	0.11	10.25-10.41	0.62	18.75-18.91	1.13
1.92- 2.08	0.12	10.42-10.58	0.63	18.92-19.08	1.14
2.09- 2.24	0.13	10.59-10.74	0.64	19.09-19.24	1.15
2.25- 2.41	0.14	10.75-10.91	0.65	19.25-19.41	1.16

2. This table shows various costs and numbers of servings. By reading the table across and down you can find the cost per serving.

   A. If a package of lunch meat costs $1.19 and serves 4 people, what is the cost per serving?

   _____

   B. If margarine costs 95¢ a pound and lasts 3 weeks, what is the cost of margarine per week?

   _____

   C. If a head of cauliflower costs 83¢ and feeds 6 people, what is the cost per serving?

   _____

Cost	Number of Servings or Weeks				
	2	3	4	5	6
0.80-0.89	0.43	0.28	0.21	0.17	0.14
0.90-0.99	0.48	0.32	0.24	0.19	0.16
1.00-1.09	0.53	0.35	0.26	0.21	0.18
1.10-1.19	0.58	0.38	0.29	0.23	0.19

3. This table shows a Saturday bus schedule. Use the table to answer the following questions.

BUS 307 SATURDAY TIMETABLE

FROM SUN VALLEY				ONE-WAY LOOP			TO SUN VALLEY			
Leave Sun Valley Shopping Center	Leave Willow Pass and Diamond	Leave Willow Pass and Fry Way	Leave Monument and Meadow Lane	Leave Oak Grove and David Avenue	Leave Bancroft and Treat Blvd.	Leave Oak Grove and David Avenue	Leave Monument and Meadow Lane	Leave Willow Pass and Fry Way	Leave Willow Pass and Diamond	Arrive Sun Valley Shopping Center
				10:22	10:27	10:32	10:36	10:43	10:47	10:55
10:55	11:03	11:08	11:15	11:22	11:27	11:32	11:36	11:43	11:47	11:55
11:55	12:03	12:08	12:15	12:22	12:27	12:32	12:36	12:43	12:47	12:53
12:53	**1:01**	**1:06**	**1:13**	**1:17**	**1:22**	**1:32**	**1:36**	**1:43**	**1:47**	**1:53**
1:53	**2:01**	**2:06**	**2:13**	**2:17**	**2:22**	**2:32**	**2:36**	**2:43**	**2:47**	**2:53**
2:53	**3:01**	**3:06**	**3:13**	**3:17**	**3:22**	**3:32**	**3:36**	**3:43**	**3:47**	**3:53**
3:53	**4:01**	**4:06**	**4:13**	**4:17**	**4:22**	**4:32**	**4:36**	**4:43**	**4:47**	**4:53**
4:53	**5:01**	**5:06**	**5:13**	**5:17**	**5:22**	**5:27**				

Light Face Figures A.M.
Dark Face Figures P.M.

A. What number bus takes passengers from Sun Valley to Oak Grove?

_____

B. How many times does the bus stop at Bancroft and Treat Blvd. on Saturday morning?

_____

C. If you worked at Sun Valley and got off work at 2 P.M., how long would you have to wait for the bus?

_____

D. If you lived near Willow Pass and Fry Way, what is the earliest you could catch a bus to Sun Valley?

_____

UNIT 4 ■ TABLES, GRAPHS, AND AVERAGES

125

4. Use the calorie table to answer the following questions.

   A. Curt had 1 serving each of meat loaf, corn, and green beans for dinner. How many calories did he have?

   _____

   B. If you wanted to diet by eating low calorie foods, which 3 foods in the table have the fewest calories?

   _____

   _____

   _____

   C. List the 3 foods you like best in the table. How many calories per serving does each have?

Food	Calories
_____	_____
_____	_____
_____	_____

Calorie Table

Item	Amount	Calories
Pork chop	3 ounces (1)	308
Meat loaf	3-ounce slice	230
Tuna	3 ounces	168
Raisins	$4\frac{1}{2}$ tablespoons	123
Orange	1	65
Peaches	$\frac{1}{2}$ cup	39
Asparagus	$\frac{1}{2}$ cup	12
Beans, green	$\frac{1}{2}$ cup	16
Corn	$\frac{1}{2}$ cup	70
Fried potatoes	20 pieces	233
Bread	1 slice	55
Bun	$\frac{1}{2}$	60
Milk	1 cup	145
Ice cream	$\frac{1}{2}$ cup	138

READING GRAPHS

A graph is a visual way to present information. The most commonly used graphs are line graphs, bar graphs, circle graphs, and picture graphs. The type of graph used depends on the type of information to be shown. To interpret a graph, carefully study the title, labels, and other given information.

Line Graphs

A *line graph* shows how numbers change, often over a period of time.

There is *horizontal* (left to right) information across the bottom of a line graph. There is *vertical* (bottom to top) information at the left of a line graph. The line of a line graph shows how the horizontal and vertical information are related.

Example 1   How many hamburgers were sold in February?

Monthly Hamburger Sales

Find February at the bottom of the graph. Move up to where the line crosses the line for February. Move from that point straight across to the vertical information at the left of the graph.

Answer: 1,000 hamburgers were sold in February.

> **HINT:** Sometimes if the intersection is not exactly on a labeled point, you must estimate the value of that intersection.

Example 2   How many hamburgers were sold in May?

Find May at the bottom of the graph. Move up to where the line crosses the line for May. This point is about halfway between the 2,000 and the 2,500 at the left of the graph.

Answer: About 2,250 hamburgers were sold in May (2,000 + 250).

Use the line graphs to answer the following questions.

1. The line graph shows how much money Donice had in her savings account at the beginning of each year from 1985 to 1989.

    A. About how much money did Donice have in her savngs account in 1986?

    _____

    B. How much money was in the savings account in 1988?

    _____

    C. In which year did Donice save the most?

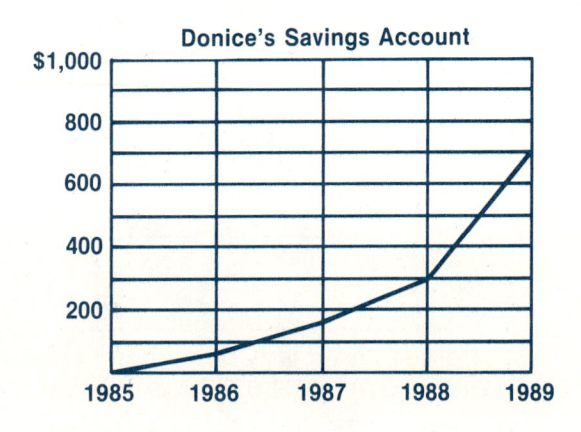

    _____

UNIT 4 ■ TABLES, GRAPHS, AND AVERAGES

127

2. The line graph shows the temperatures every hour from 8 A.M. to 3 P.M. for 1 day.

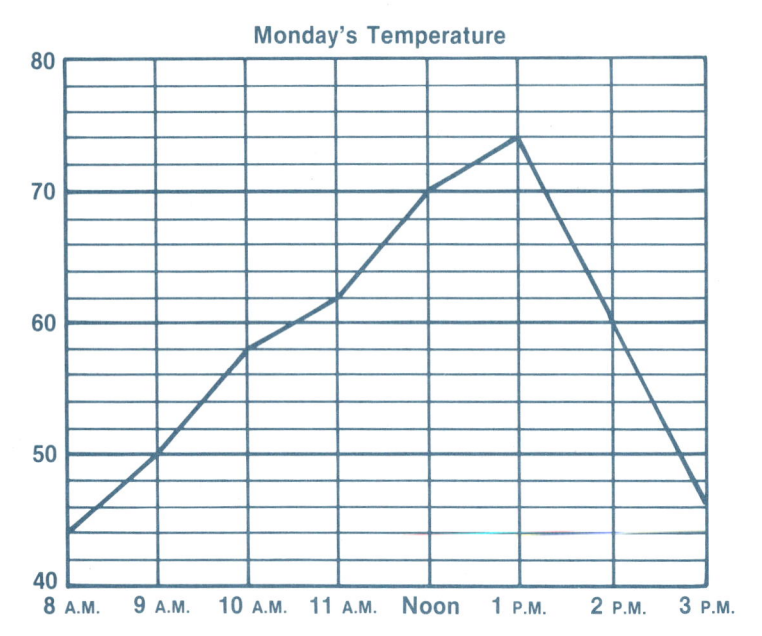

Monday's Temperature

A. What was the highest temperature Monday?
_____

B. When did the highest temperature occur?
_____

C. What was the temperature change from 9 A.M. until noon?
_____

D. What was the difference between the highest and lowest temperatures?
_____

3. The line graph shows both the total number of people in the labor force *and* the number of women in the labor force. Note that the vertical information refers to millions.

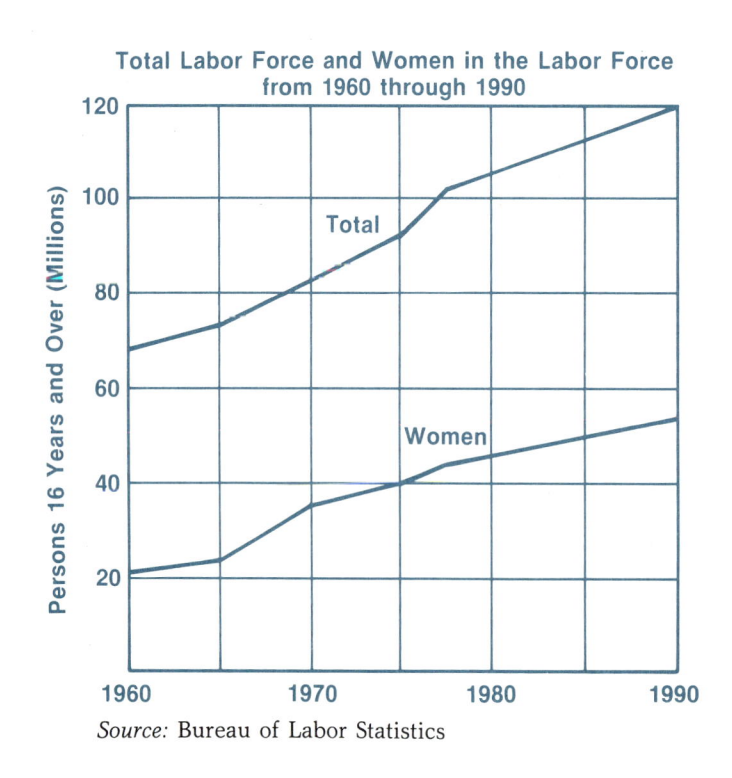

Total Labor Force and Women in the Labor Force from 1960 through 1990

Source: Bureau of Labor Statistics

A. How many women were in the labor force in 1975?
_____

B. During what 5-year period did the number of women workers increase the most?
_____

C. In 1985, approximately what percent of the work force was made up of women?
_____

D. The total labor force increases from 1970 to 1990 by what amount?
_____

128   UNIT 4 ■ TABLES, GRAPHS, AND AVERAGES

4. Drivers 15 to 19 years old have very high traffic accident, injury, and conviction rates. Traffic accidents are the leading cause of deaths for teenagers.

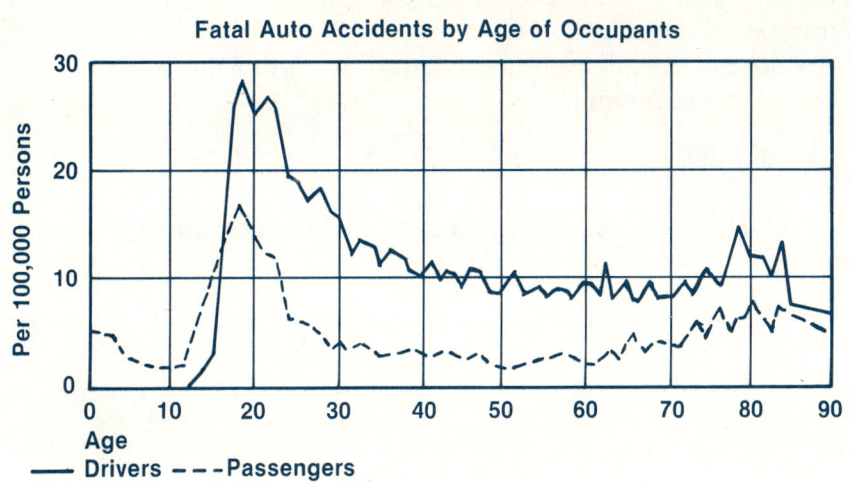

Fatal Auto Accidents by Age of Occupants

Per 100,000 Persons (vertical axis: 0, 10, 20, 30)
Age (horizontal axis: 0, 10, 20, 30, 40, 50, 60, 70, 80, 90)
—— Drivers - - - Passengers

Source: Insurance Institute of Highway Safety

A. What group of people is represented by the dashed line? _____

B. What do the vertical numbers, 0, 10, 20, and 30 represent? _____

C. At what age do the most fatal accidents occur? _____

D. Approximately how many teen passengers died at age 17? _____

E. At what age are fatal accidents lowest for the legal driver? _____

UNIT 4 ■ TABLES, GRAPHS, AND AVERAGES

Bar Graphs

A *bar graph* shows a comparison of quantities. As in line graphs, bar graphs have information across the bottom and at the left of the graph. Vertical or horizontal bars are used to compare the information.

If the bars are vertical, read directly across from the top of the bar to the information on the side.

If the bars are horizontal, read down from the end of the bar to the information at the bottom.

> **HINT:** Use a ruler or piece of paper to line up the bar with the information.

Example 1   What percent of the students chose swimming as their favorite sport?

Vertical Bar Graph

Horizontal Bar Graph

Answer: 20% of the students chose swimming.

When the end of the bar falls between two numbers, estimate the amount.

Example 2   What percent of the students chose football as their favorite sport?

The end of the bar for football is between 50% and 60%. So approximately 55% of the students chose football.

Use the bar graphs to answer the following questions.

1. Mr. Campbell surveyed some of his students regarding the school lunch food.

Students' Reaction to Lunch

(Feeling About Food vs. Number of Students)
- Strongly Dislike: 50
- Dislike: 20
- Like: 10
- Love: 5

A. How many students loved the food? _____

B. How do most students feel about the food? _____

C. How many students did Mr. Campbell survey? _____

2. Four students compared their test scores by using a bar graph.

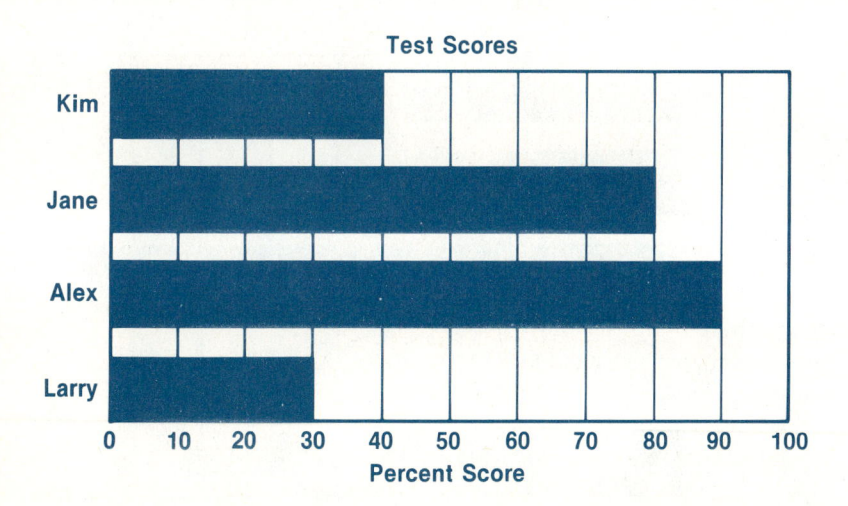

Test Scores

- Kim: 40
- Jane: 80
- Alex: 90
- Larry: 30

A. Which student had the highest score? _____

B. If you combine Kim's and Larry's scores, will the combination be as high as Alex's score? _____

C. How much higher was Alex's score than Jane's score? _____

D. What was Jane's score? _____

UNIT 4 ■ TABLES, GRAPHS, AND AVERAGES

3. The bar graph shows the distances from San Francisco to 6 cities.

   A. Which city is closest to San Francisco?

   _____

   B. Approximately how many miles is Miami from San Francisco?

   _____

   C. What is the approximate minimum number of miles you would travel, round trip, from San Francisco to Houston?

   _____

   D. Cincinnati is approximately how many miles closer to San Francisco than Washington D.C. is to San Francisco?

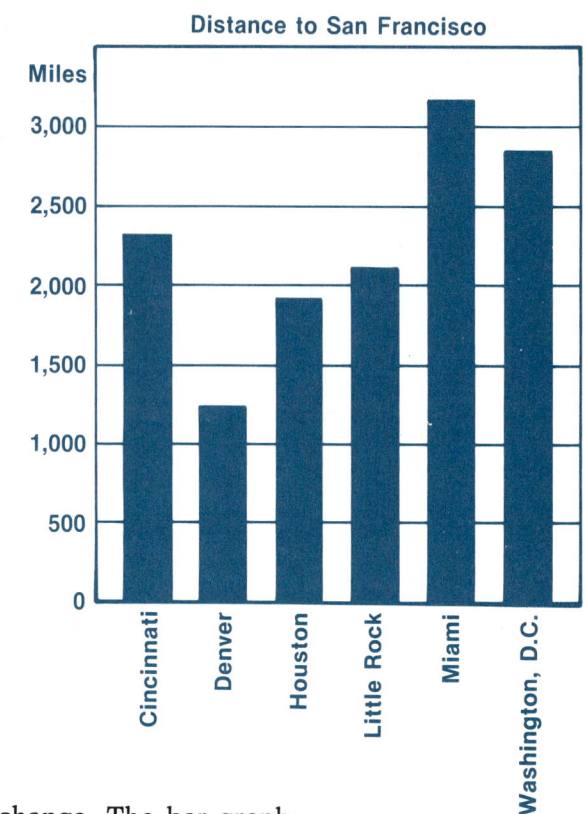

   _____

4. The job market changes as companies change. The bar graph shows the numbers of jobs that will be available from 1980 to 1990.

   A. Of the jobs shown, in which area will there be the most openings?

   _____

   B. Approximately how many job openings will be available in sales?

   _____

   C. How many more job openings will be available to service workers than to professional and technical workers?

   _____

   D. What percent of the total job openings shown are job openings for nonfarm laborers? (Round to the nearest percent.)

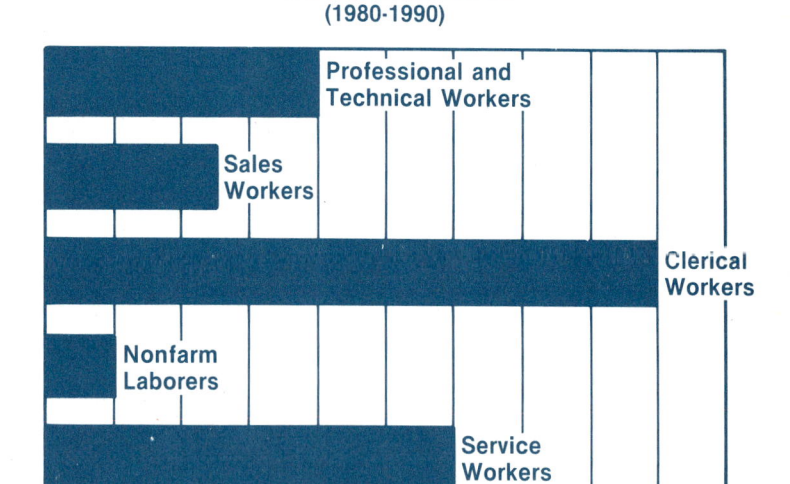

   _____

Source: Bureau of Labor Statistics, *Occupational Outlook Handbook,* 1980-81 ed.

132   UNIT 4 ■ TABLES, GRAPHS, AND AVERAGES

Circle Graphs

A circle graph shows how the parts of something compare in size with each other and with the whole. The total parts of the circle will always equal 100%.

Example What percent of the records sold were jazz records?

Find the jazz category on the circle graph and read the percent.

Answer: 15% of the records sold were jazz records.

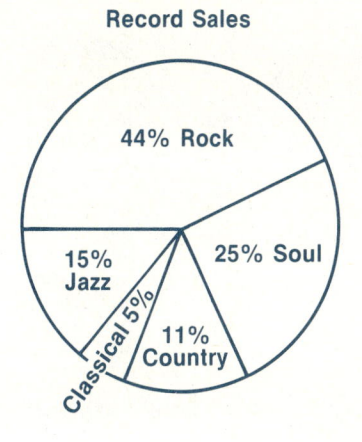

Record Sales: 44% Rock, 25% Soul, 11% Country, 5% Classical, 15% Jazz

Use the circle graphs to answer the following questions.

1. The circle graph shows the types of jobs held by women.

   A. Of every 100 working women how many women work as clerical workers?

   _____

   Women in the Work Force Today (40 Million Working Women): Professionals 16%, Service Workers 21%, Clerical Workers 35%, Sales Workers 7%, Managers 6%, Factory Workers 15%

   B. Approximately how many women work as clerical workers?

   _____

   C. Write a ratio comparing the number of women service workers to women sales workers.

   _____

   D. The number of professional women is how much greater than the number of women factory workers?

   _____

UNIT 4 ■ TABLES, GRAPHS, AND AVERAGES

2. The two circle graphs show how the federal government spent its money for two different years. (Note that the two graphs represent different amounts of total money spent.)

Federal Government Spending

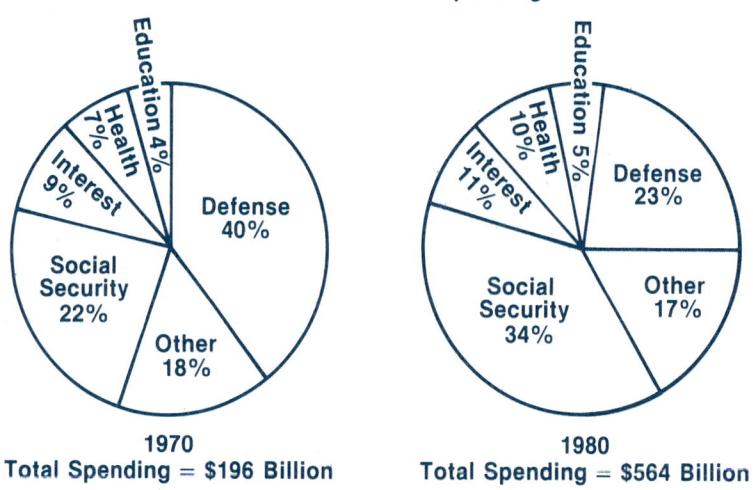

1970
Total Spending = $196 Billion

- Defense 40%
- Social Security 22%
- Other 18%
- Interest 9%
- Health 7%
- Education 4%

1980
Total Spending = $564 Billion

- Defense 23%
- Other 17%
- Social Security 34%
- Interest 11%
- Health 10%
- Education 5%

Source: U.S. Bureau of the Census, *Statistical Abstracts of the United States, 1980.*

A. How much money was spent by the federal government in 1980? _____

B. In 1970, of every $100 spent, how much money was spent on education? _____

C. In 1980, of every $100 spent, how much money was spent on education? _____

D. What was the dollar amount the federal government spent for defense in 1970? _____

E. What was the dollar amount the federal government spent for defense in 1980? _____

F. What was the dollar amount the federal government spent for education in 1970? _____

G. What was the dollar amount the federal government spent for education in 1980? _____

3. The circle graph shows the income of the federal government.

Federal Government Income, 1979
Total Income: $530 Billion

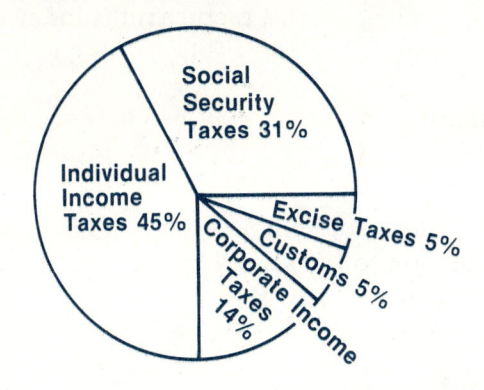

Social Security Taxes 31%
Individual Income Taxes 45%
Excise Taxes 5%
Customs 5%
Corporate Income Taxes 14%

Source: U.S. Bureau of the Census, *Statistical Abstracts of the United States, 1980.*

A. From what two sources did the federal government receive the most income?

_____

_____

B. What is the dollar amount the federal government received from individual income taxes?

_____

C. How much money did the government receive from Social Security?

_____

Picture Graphs

Sometimes newspapers, magazines, and television will represent information in a picture graph. A *picture graph* uses picture units to represent number values. Most picture graphs show approximate numbers rather than exact figures.

Example 1   A committee made a survey of activities in which students participated during the summer. How many students went hiking?

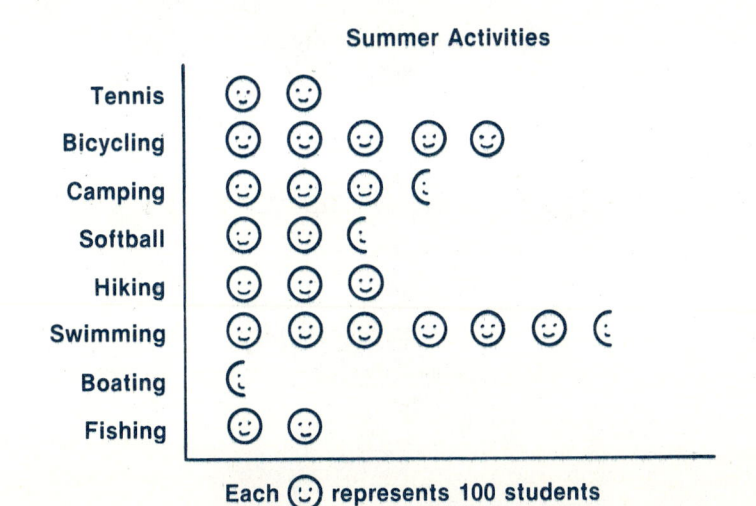

Summer Activities

Tennis	☺ ☺
Bicycling	☺ ☺ ☺ ☺ ☺
Camping	☺ ☺ ☺ ☹
Softball	☺ ☺ ☹
Hiking	☺ ☺ ☺
Swimming	☺ ☺ ☺ ☺ ☺ ☺ ☹
Boating	☹
Fishing	☺ ☺

Each ☺ represents 100 students

Find hiking at the left of the graph. There are 3 picture units for hiking and each picture represents 100 students: 3 × 100 students = 300 students.

Answer: 300 students went hiking.

UNIT 4 ■ TABLES, GRAPHS, AND AVERAGES

A partial picture in a picture graph indicates that there are not enough numbers to make the next unit complete.

Example 2   In the picture graph on page 135, how many students went swimming?

Find swimming at the left of the graph. There are $6\frac{1}{2}$ picture units for swimming and each picture represents 100 students: $6\frac{1}{2} \times 100$ students = 650 students.

Answer: 650 students went swimming.

Teachers in Hometown School District compared their student class size to that of teachers in other nearby districts. They presented their information to the school board using the picture graph below. Use the graph to answer the questions that follow.

Comparison of Student Class Size

Hometown District	☹	☹	☹	☹	☹	☹
Central District	☺	☺	☺	☻		
Union District	☺	☺	☺			
Middletown District	😐	😐	😐	😐	🫤	
Sunnyside District	😐	😐	😐	😐		

Each face represents 6 students

1. How does the graph illustrate the feelings of the students and teachers in the Hometown District?

   _____

   _____

2. What does half a picture unit represent? _____

3. How many students are in the classrooms at Middletown District? _____

136   UNIT 4 ■ TABLES, GRAPHS, AND AVERAGES

MAKING GRAPHS FROM GIVEN INFORMATION

It is easy to transfer a list of facts to graph form. A graph helps present information clearly and quickly. Graphs also allow the viewer to draw conclusions that may not be obvious from a list of written or verbal information.

Line Graphs

Follow these steps to make a line graph:

1. Choose a title for the graph.
2. Label the horizontal and vertical information. (What facts are being compared?)
3. Plot points on the graph to show where two sets of facts meet and intersect.
4. Connect the points.

Draw line graphs and answer the questions that follow.

1. Amy is saving money to buy a VCR that costs $360. She works in a video store after school and makes $60 per week. She figures she can use half her money for normal expenses and save the other half of her paychecks. She begins saving on April 3. After 3 weeks, she decides she can save $10 more each week. Use the grid below to draw a graph of Amy's savings.

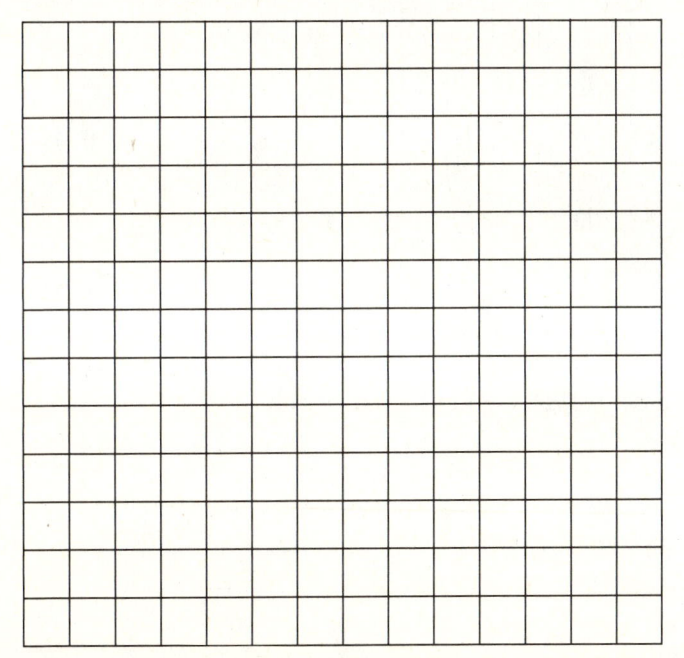

   A. How much money will Amy save by May 1? _____

   B. How much money will Amy have saved in 6 weeks? _____

   C. Will Amy have enough saved by June 1 to buy the VCR? If not, how much will she need? _____

UNIT 4 ■ TABLES, GRAPHS, AND AVERAGES

137

2. Kirk would like to walk out of the desert before his canteen runs out of water. His canteen has 30 ounces of water and he must drink 3 ounces per hour. He can walk 2.5 miles per hour. Use the grid below to draw a graph that shows how far he can walk and how much water he uses. (Use miles for the vertical information and ounces of water for the horizontal information.)

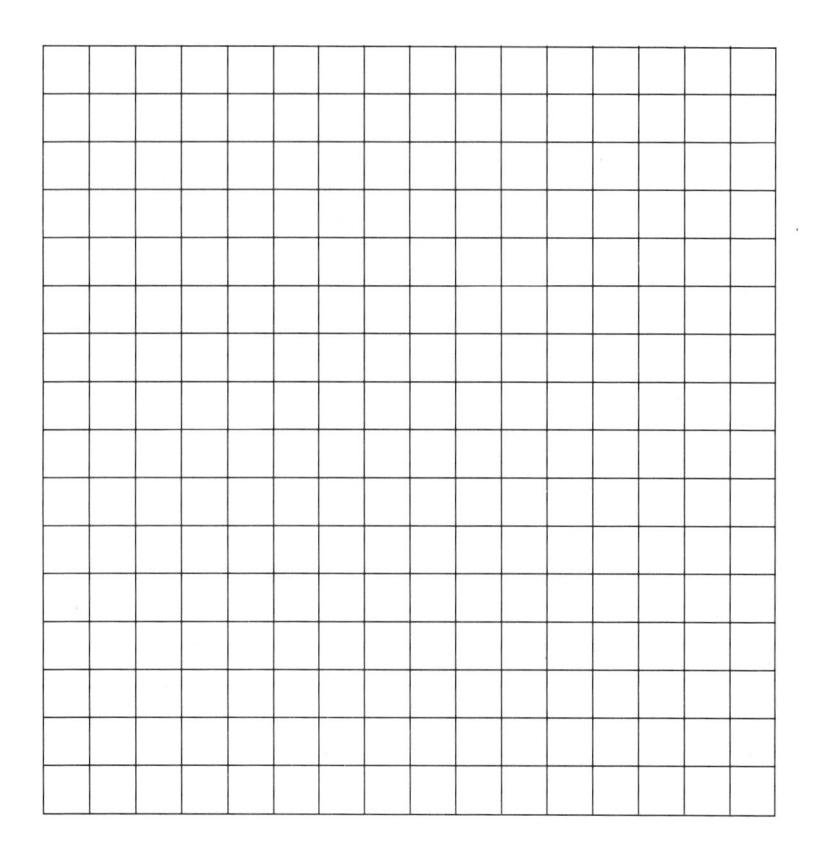

A. How many miles will Kirk have walked when he has drunk 12 ounces of water? _____

B. How many miles will he have walked when half his water is gone? _____

C. How much water will he have after he has walked 15 miles? _____

D. If Kirk is in the middle of a 40-mile wide desert, do the buzzards win? _____

Bar Graphs

Follow these steps to make a bar graph:

1. Choose a title for the graph.
2. Decide if the bars are to be vertical or horizontal.
3. Space the items to be compared on the left side if the bars are to be horizontal. Space them on the bottom if the bars are to be vertical.
4. Label the horizontal and vertical information.
5. Mark the data on the graph and draw the bars.

1. Use the space below to draw a bar graph that illustrates the differences in lunch selections between male and female students. Use the following data.

   35 male students and 35 female students selected pizza.
   35 female students and 5 male students selected salads.
   10 male and 45 female students selected cheese zombies.
   20 female and 25 male students selected hot dogs.
   15 male and 10 female students selected burritos.
   15 female and 10 male students selected sandwiches.

   HINT: **For each food choice use two bars, one for male choices and one for female choices. Shade in the bars for male choices differently than the bars for female choices. Indicate which shading is for male and which is for female.**

UNIT 4 ■ TABLES, GRAPHS, AND AVERAGES

2. Use the space below to draw a bar graph that illustrates the following information. These are the 10 occupations with the largest number of new jobs (in thousands), as estimated for the years 1985 to 1995. The source for this information is the California State Employment Development Department, SF Chronicle; September 1988.

   Retail salespeople: 113.8
   Cashiers: 88.8
   Waiters and waitresses: 78.0
   General managers and executives: 77.2
   General office clerks: 70.2
   Registered nurses: 67.1
   Secretaries: 58.9
   Janitors, cleaners: 52.1
   Auditors and accountants: 47.9

   Place the occupations in alphabetical order on your bar graph and include the source at the bottom of the graph.

Circle Graphs

Follow these steps to make a circle graph:

1. Choose a title for your graph.
2. Draw a circle.
3. Estimate the fractional size of the circle for each category you need to show.
4. Mark your circle by drawing fractional sections to show the comparative sizes of the data.
5. Label each fractional section of the circle.

> **HINT:** To estimate a fractional part of a circle, think of the face of a clock: $\frac{1}{3}$ would be 4:00, $\frac{1}{4}$ would be 3:00, $\frac{1}{6}$ would be 2:00, $\frac{1}{12}$ would be 1:00.

1. Tamara's parents want her to wash the car and mow the lawn Saturday. She wants to show them that she doesn't have time. (The combined chores would probably take her $2\frac{1}{2}$ hours to complete.) Draw a 24-hour circle graph showing her parents that she doesn't have time Saturday for household chores. Use the following information.

    She sleeps 8 hours.
    She works at the local theater from 1 P.M. to 9 P.M.
    Her commute to work and home takes $\frac{1}{2}$ hour each way.
    She needs to spend 4 hours at the library researching a term paper.
    She needs the remaining part of the day to eat meals, shower, iron her uniform, and curl her hair (personal time).

2. Use the space below to draw *two* circle graphs to illustrate the following information. The source of the information is the California State Employment Development Department, SF Chronicle; September 1988.

   A. National educational requirements of new jobs:

   > At least some college: 51%
   > Less than 4 years of high school: 14%
   > High school graduates: 35%

   B. Education of the California class of 1987:

   > Attending 2- and 4-year colleges: 38%
   > High school graduates: 29%
   > Less than 4 years of high school: 33%

Picture Graphs

Follow these steps to make a picture graph:
1. Choose a title for your graph.
2. Choose a picture unit.
3. List the items to be graphed on the left side of a vertical line.
4. Draw the correct number of whole and partial pictures for each item listed.

A teacher said that students could listen to the radio in class if 75% of the students agreed on the same type of music. Use the space below to draw a picture graph that illustrates the following student preferences. Then answer the questions that follow.

10 students preferred western music.
22 students preferred rock music.
3 students preferred classical music.
0 students preferred swing music.

1. The student's first choice was preferred by what percent of the students? _____

2. Were the students able to listen to the radio in class? _____

UNIT 4 ■ TABLES, GRAPHS, AND AVERAGES

PRACTICE WITH TABLES AND GRAPHS

Use the tables and graphs to answer the questions in this practice section.

1. Study the example. Then determine the federal income tax owed in the following situations by reading the tax table to the right.

Example Mr. and Mrs. Brown are filing a joint return. Their taxable income on line 37 of Form 1040 is $25,325. First, they find the $25,300-25,350 income line. Next, they find the column for married filing jointly and read down the column. The amount shown where the income line and filing status column meet is $3,470. This is the tax amount they must write on line 38 of their return.

At least	But less than	Single	Married filing jointly*	Married filing separately	Head of a household
			Your tax is—		
25,200	25,250	4,406	3,448	5,468	4,075
25,250	25,300	4,419	3,459	5,487	4,087
25,300	25,350	4,432	**3,470**	5,506	4,099
25,350	25,400	4,446	3,481	5,525	4,112

A. Mr. and Mrs. France had a combined income of $12,568. If they file jointly, how much do they owe?

B. Maurie Schumacher had an income of $11,398 and is unmarried. How much does he owe?

C. Claire Brown became a widow in the winter of last year. Her husband's and her combined income was $9,917. How much federal income tax must she pay?

D. For which filing category is the income tax highest?

If line 37 (taxable income) is—		And you are—				If line 37 (taxable income) is—		Any you are—			
At least	But less than	Single	Married filing jointly*	Married filing separately	Head of a household	At least	But less than	Single	Married filing jointly*	Married filing separately	Head of a household
			Your tax is—						Your tax is—		
8,000						**11,000**					
8,000	8,050	729	500	842	663	11,000	11,050	1,198	916	1,375	1,132
8,050	8,100	737	506	850	670	11,050	11,100	1,206	923	1,386	1,141
8,100	8,150	744	512	858	677	11,100	11,150	1,214	930	1,397	1,149
8,150	8,200	752	518	866	684	11,150	11,200	1,222	937	1,408	1,158
8,200	8,250	759	524	874	691	11,200	11,250	1,230	944	1,429	1,166
8,250	8,300	767	531	882	698	11,250	11,300	1,238	951	1,430	1,175
8,300	8,350	774	538	890	705	11,300	11,350	1,246	958	1,441	1,183
8,350	8,400	782	545	898	712	11,350	11,400	1,254	965	1,452	1,192
8,400	8,450	789	552	906	719	11,400	11,450	1,262	972	1,463	1,200
8,450	8,500	797	559	914	726	11,450	11,500	1,270	979	1,474	1,209
8,500	8,550	804	566	922	733	11,500	11,550	1,278	986	1,485	1,217
8,550	8,600	812	573	930	740	11,550	11,600	1,286	993	1,496	1,226
8,600	8,650	819	580	938	747	11,600	11,650	1,294	1,000	1,507	1,234
8,650	8,700	827	587	947	754	11,650	11,700	1,302	1,007	1,518	1,243
8,700	8,750	834	594	956	761	11,700	11,750	1,311	1,014	1,529	1,251
8,750	8,800	842	601	965	768	11,750	11,800	1,320	1,021	1,540	1,260
8,800	8,850	849	608	974	775	11,800	11,850	1,329	1,028	1,551	1,268
8,850	8,900	857	615	983	782	11,850	11,900	1,338	1,035	1,562	1,277
8,900	8,950	864	622	992	789	11,900	11,950	1,347	1,042	1,573	1,285
8,950	9,000	872	629	1,001	796	11,950	12,000	1,356	1,049	1,584	1,294
9,000						**12,000**					
9,000	9,050	879	636	1,010	803	12,000	12,050	1,365	1,056	1,595	1,302
9,050	9,100	887	643	1,019	810	12,050	12,100	1,374	1,063	1,606	1,311
9,100	9,150	894	650	1,028	817	12,100	12,150	1,383	1,070	1,617	1,319
9,150	9,200	902	657	1,037	824	12,150	12,200	1,392	1,077	1,628	1,328
9,200	9,250	910	664	1,046	831	12,200	12,250	1,401	1,084	1,639	1,336
9,250	9,300	918	671	1,055	838	12,250	12,300	1,410	1,091	1,650	1,345
9,300	9,350	926	678	1,064	845	12,300	12,350	1,419	1,098	1,661	1,353
9,350	9,400	934	685	1,073	852	12,350	12,400	1,428	1,105	1,672	1,362
9,400	9,450	942	692	1,082	860	12,400	12,450	1,437	1,112	1,683	1,370
9,450	9,500	950	699	1,091	869	12,450	12,500	1,446	1,119	1,694	1,379
9,500	9,550	958	706	1,100	877	12,500	12,550	1,455	1,126	1,705	1,387
9,550	9,600	966	713	1,109	886	12,550	12,600	1,464	1,133	1,716	1,396
9,600	9,650	974	720	1,118	894	12,600	12,650	1,473	1,140	1,727	1,404
9,650	9,700	982	727	1,127	903	12,650	12,700	1,482	1,147	1,738	1,413
9,700	9,750	990	734	1,136	911	12,700	12,750	1,491	1,154	1,749	1,421
9,750	9,800	998	741	1,145	920	12,750	12,800	1,500	1,161	1,760	1,430
9,800	9,850	1,006	748	1,154	928	12,800	12,850	1,509	1,168	1,771	1,439
9,850	9,900	1,014	755	1,163	937	12,850	12,900	1,518	1,176	1,782	1,448
9,900	9,950	1,022	762	1,172	945	12,900	12,950	1,527	1,184	1,793	1,457
9,950	10,000	1,030	769	1,181	954	12,950	13,000	1,536	1,192	1,804	1,466

*This column must also be used by a qualifying widow(er).

2. Use the line graph to answer the following questions.

   A. During which month was there the greatest temperature change?

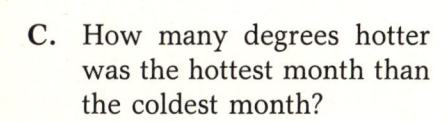

   _____

   B. During which two months did the temperature remain the same?

   _____

   _____

   C. How many degrees hotter was the hottest month than the coldest month?

   _____

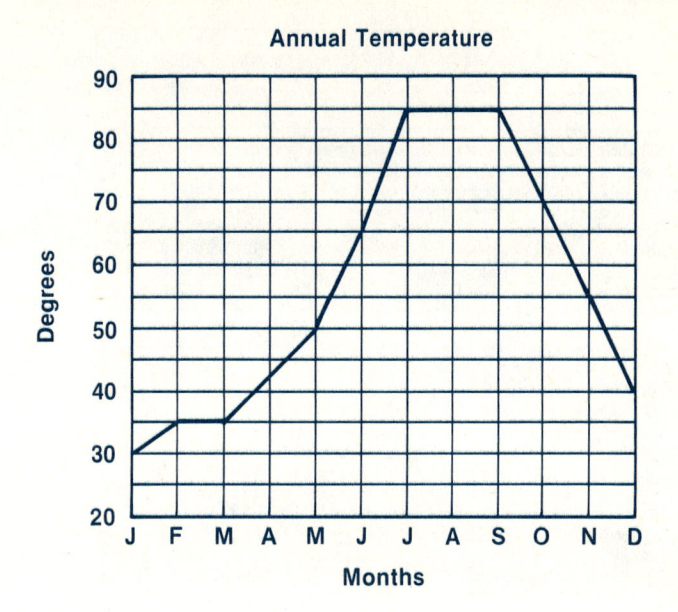

Annual Temperature

3. Use the bar graph to answer the following questions.

Percentage of Population Increase

1940-1945: 1.2
1945-1950: 1.6
1950-1955: 1.7
1955-1960: 1.7
1960-1965: 1.5
1965-1970: 1.1
1970-1975: 0.8
1975-1980: 0.8
1980-1985: 0.9
1985-1990: 0.9

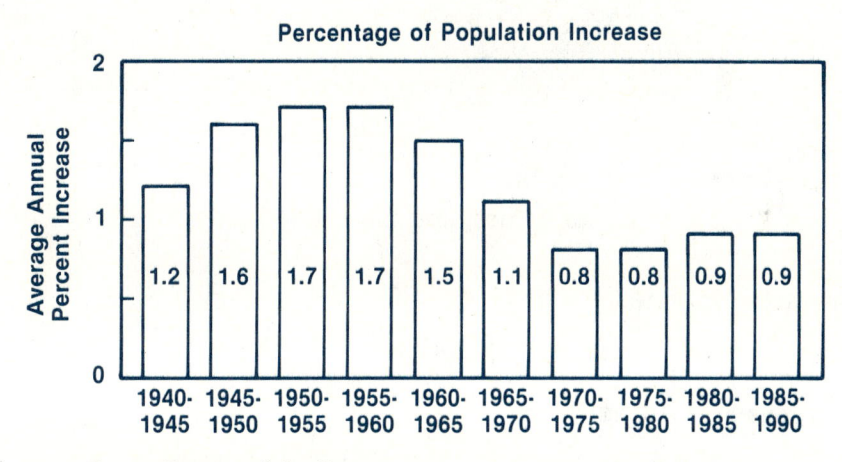

Source: Bureau of the Census

   A. What do the decimal numbers represent? _____

   B. During which 10-year period will the school have the greatest increase in kindergarten enrollment? _____

   C. If the population of a town was 26,000 in 1985, and its 5-year percent increase in population corresponded to the chart above, what would its population be in 1990? _____

UNIT 4 ■ TABLES, GRAPHS, AND AVERAGES

4. The table on the back of a pattern envelope shows information necessary for sewing the garment.

MISSES' SHIRT, PANTS AND SHORTS

Metric conversion chart given on enclosed direction sheet.

Extra fabric needed to match plaids, stripes, one-way designs. Use nap yardage and nap layouts for one-way design fabrics. Not suitable for obvious diagonal fabrics.

STANDARD BODY MEASURE-MENTS	Bust	30½	31½	32½	34	36	Ins
	Waist	23	24	25	26½	28	"
	Hip 9" below waist	32½	33½	34½	36	38	"
	Back—neck to waist	15½	15¾	16	16¼	16½	"

Fabric required	Sizes	6	8	10	12	14	
Shirt							
35" or 36" without nap		2⅛	2¼	2¼	2⅜	2½	Yds
44" or 45" without nap		1½	1½	1⅝	1¾	1¾	"
58" or 60" without nap		1⅛	1⅛	1¼	1¼	1¼	"
Interfacing — woven or non-woven, fusible or non-fusible.							
25", 32", 35" or 36"		1½	1½	1½	1⅝	1⅝	"
Shorts							
35" or 36" without nap		1⅛	1¼	1¼	1¼	1⅜	"
44" or 45" without nap		⅞	1	1	1	1⅛	"
58" or 60" with or without nap		¾	¾	¾	⅞	⅞	"
Pants							
35" or 36" without nap		2⅝	2⅝	2⅝	2¾	2¾	"
44" or 45" with or without nap		2¼	2¼	2¼	2½	2⅝	"
58" or 60" with or without nap		1½	1½	1⅝	1⅝	1⅝	"

Garment Measurements
Finished back length of shirt   13½ 13¾ 14   14¼ 14½ Ins
Finished length at side seam from waistline seam line of:
shorts   12   12¼ 12½ 12¾ 13   "
pants   41½ 41¾ 42   42½ 42¼   "
Bottom width of pants leg   19   19½ 20   20¾ 21¾   "

Sewing notions — Thread, seam binding or stretch lace. Shirt: Six ½" buttons. Pants or Shorts: 7" zipper, two ½" buttons (opt).

7632 Misses' Shirt, Pants, and Shorts

A. What size pattern would you buy for a person with a bust of $32\frac{1}{4}$ inches, a waist of 25 inches, and hips of 34 inches? _____

B. How much fabric would you need for a shirt, size 6, 45-inch fabric? _____

C. What notions are needed for the shirt? _____

D. What notions are needed for the pants? _____

E. What is the bottom width of the pants leg for size 8? _____

F. When might you need extra fabric? _____

G. What is the waist measurement for a size 12 pattern? _____

UNIT 4 ■ TABLES, GRAPHS, AND AVERAGES

5. Computer use in the classroom is increasing every year. The number of computers grew from 250,000 in 1983 to 1 million in 1985.

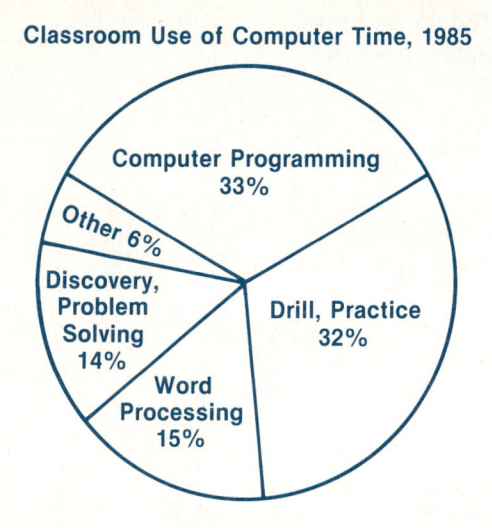

Classroom Use of Computer Time, 1985
- Computer Programming 33%
- Other 6%
- Discovery, Problem Solving 14%
- Word Processing 15%
- Drill, Practice 32%

A. Write a ratio comparing the time spent on drill and practice to the time spent on word processing.

B. By what percent did the number of computers in the classroom increase from 1983 to 1985?

C. Computer programming and word processing make up what percent of classroom use of computer time?

6. Ordering items from a catalog is sometimes convenient or necessary. Generally, the procedure for ordering is as follows:
   - Carefully read all the information regarding the product.
   - Select the item, size, color, and quantity.
   - Decide whether to have the product sent to your home or to the nearest catalog store.
   - Figure total cost, including tax and shipping.

To practice ordering, pretend that you live in a city that has a 5% sales tax, and that you have just purchased a 1972 Corvette. You want to order a black, polypropylene Complete Set carpet replacement for the car. You live in Zone 2, and will have the kit mailed directly to your home. Study the tables and answer the following questions.

Custom-Fit Replacement Carpets for Corvettes, Imported Cars, Mini-Pickups, and 4-Wheel Drive Vehicles. Your choice of two tufted piles: polypropylene loop pile or nylon cut pile. Car carpets have ¼-inch-thick polyurethane foam underpadding, others have jute. Vinyl-bound edges. Driver's side heel-protector pad. Contour-sewn pieces. Front Set—covers left and right front floors (excluding transmission tunnel) Center Set—covers transmission tunnel. Complete Set—covers all sections of vehicle which were originally carpeted. Order from chart below. Polypropylene colors: 04 Burgundy; 36 Dark Blue; 54 Brown; 82 Black; Nylon colors: 03 Red; 35 Blue; 54 Brown; 82 Black. State color number-and-name.

4WD Pickup	Set	Lbs	Polypropylene	Price	Nylon Cut Pile	Price
Jeep® CJ5, CJ7	Front	12	A 976-3608 FH	74.99	A 976-3616 FH	84.99
	Complete Set	25	A 976-3624 FH	119.99	A 976-3632 FH	134.99
Cherokee, Wagoneer 84-6	Front	12	A 976-3640 FH	74.99	A 976-3657 FH	84.99
	Complete Set	25	A 976-3665 FH	149.99	A 976-3673 FH	189.99
Chevy Suburban 1976-86	Front	12	A 976-3681 FH	49.99	A 976-3699 FH	59.99
	Complete Set	25	A 976-3707 FH	164.99	A 976-3715 FH	199.99
Blazer/Jimmy, Bronco	Front	12	A 970-2044 FH	74.99	A 970-2051 FH	84.99
	Complete Set	25	A 970-2069 FH	149.99	A 970-2077 FH	169.99
Mini Pickup	Front Center	10	A 987-5857 FH	44.99	A 987-5865 FH	54.99
Car	**Set**	**Lbs**	**Polypropylene**	**Price**	**Nylon Cut Pile**	**Price**
Corvette 54-83 63-83	Front and Center	10	A 967-5758 FH	49.99	A 967-5766 FH	59.99
	Complete Set	13	A 963-6735 FH	179.99	A 964-8474 FH	199.99
Datsun Cars	Front and Center	10	A 968-1665 FH	44.99	A 986-5726 FH	59.99
Datsun Cars	Front, Rear and Center	12	A 970-2002 FH	84.99	A 970-2010 FH	94.99
Datsun 240, 260 and 280Z	Complete Set	15	A 970-2028 FH	94.99	A 970-2036 FH	114.99
Honda Civic and Accord	Front and Center	10	A 968-2101 FH	44.99	A 986-9975 FH	54.99

FOR HOME DELIVERY

Delivery Weight of Your Order in Lbs	Zone I (Local) Cat. Depts. in Distribution Center Cities	Zone II All Other Catalog Depts.
0 - 0.5 lb	$0.97	$1.26
0.6 - 1 lb	1.38	1.71
1.1 - 2 lbs	1.69	2.27
2.1 - 3 lbs	1.76	2.49
3.1 - 4 lbs	1.88	2.81
4.1 - 5 lbs	1.95	3.04
5.1 - 6 lbs	2.01	3.28
6.1 - 7 lbs	2.08	3.55
7.1 - 8 lbs	2.14	3.78
8.1 - 9 lbs	2.18	3.93
9.1 - 10 lbs	2.23	4.01
10.1 - 11 lbs	2.30	4.09
11.1 - 12 lbs	2.34	4.17
12.1 - 13 lbs	2.40	4.24
13.1 - 14 lbs	2.44	4.35
14.1 - 15 lbs	2.50	4.48
15.1 - 16 lbs	2.55	4.63
16.1 - 17 lbs	2.61	4.77

A. What is the catalogue number for your order?

B. What is the color number for your order?

C. What is the total cost of this catalog order, including sales tax and shipping cost?

UNIT 4 ■ TABLES, GRAPHS, AND AVERAGES

FINDING AVERAGES: MEAN, MEDIAN, MODE

The mean, median, and mode are measures that tell something about a group of numbers. The illustration and table below show the use of each of these words.

Heights of Basketball Players

Marilyn	5' 3"
Bud	5' 7"
Connie	5'10"
Wayne	5'11"
Ramona	6' 1"
Stan	6' 2"
Sonny	6' 2"
Doug	6' 5"
Willie	6' 7"
Total	54'

Heights of Basketball Players

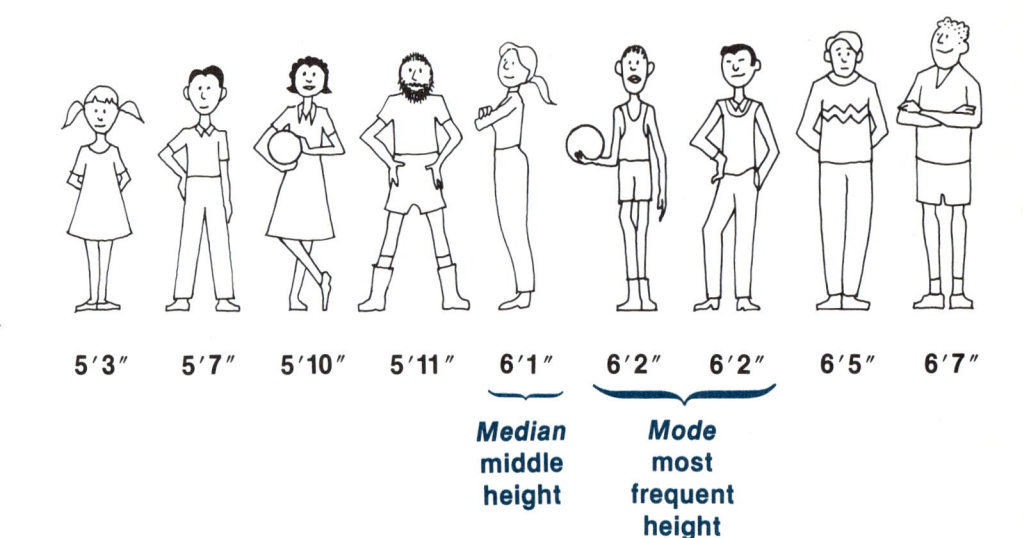

5'3" 5'7" 5'10" 5'11" 6'1" 6'2" 6'2" 6'5" 6'7"

Median middle height

Mode most frequent height

Mean: A number obtained by adding a series of numbers and dividing that total by the number of items in the series.

The sum of the players' heights is 54' and there are 9 players. Divide: 54' ÷ 9 = 6'. The mean height is 6'.

Median: The middle number in a series of numbers.

The median height is 6'1".

Mode: The most frequent number in a series of numbers.

The mode height is 6'2".

Use the table below to answer the following questions.

1. What is the mean salary? _____

2. What is the median salary? _____

3. What is the mode salary? _____

Student's Part-Time Monthly Salary

Alicia	$105
Elaine	$145
Lester	$176
Suzie	$176
Paul	$181
Prentice	$200
Stephanie	$220
Julio	$255
Rachel	$306

The word *average* is used interchangeably with the word *mean*. In this text, the word *average* will refer to the mean of information.

To find the average, first add all the items; then divide the total by the number of items.

Example On Monday the high temperature was 58°; on Tuesday it was 62°; on Wednesday it was 59°; on Thursday it was 56°; on Friday it was 53°; on Saturday it was 57°; on Sunday it was 61°. What was the average high temperature for the week?

Add the temperatures: Divide by 7 (the number of items):

```
   58°              58°
   62°          7)406°
   59°             35
   56°             ──
   53°             56
   57°             56
 + 61°             ──
  ────              0
  406°
```

Solve the following problems.

1. Pat and Lauren backpacked in Yosemite last summer. What was their average daily hike if they walked 2.5 kilometers, 3 kilometers, 2.5 kilometers, 3 kilometers, and 1.5 kilometers? _____

2. Emi worked $6\frac{1}{2}$ hours Monday, $7\frac{3}{4}$ hours Tuesday, 5 hours Wednesday, $4\frac{1}{2}$ hours Thursday, and $6\frac{1}{4}$ hours Friday. How many hours did she average per day? _____

UNIT 4 ■ TABLES, GRAPHS, AND AVERAGES

3. Jerry's gasoline bill was $82.40 in July, $91.85 in August, $20.62 in September, and $64.85 in October. What was his average bill for those months?

_____

4. The local used-car lot sold 4 cars yesterday. The cars were sold for $4,065, $3,983, $4,117, and $5,267. What was the average selling price of the cars?

_____

5. The Donovan family traveled to Arizona last summer. They drove 487 miles Monday, 392 miles Tuesday, and 456 miles Saturday. How many miles did they average per day?

_____

6. Beverly decided to figure how many miles per gallon her car was averaging. When she bought gas Thursday, the odometer read 29,853 miles. The next time she bought gas, she bought 13 gallons to fill the tank, and the new odometer reading was 30,074 miles. How many miles did she average per gallon of gas?

_____

7. The rainfall last year was 0.3 inches in September, 1.2 inches in October, 2.6 inches in November, and 3.5 inches in December. What was the average monthly rainfall in that time period? _____

8. Paula drove 382.5 miles in $7\frac{1}{2}$ hours. How many miles per hour did she average? _____

9. Shauna is trying to gain weight. Her calorie intake was 2,355 one day, 2,015 another day, 2,100 another day, and 1,950 another day. What was her average daily calorie intake? _____

10. An airplane left San Francisco at 1:15 P.M. and reached its destination, 1,309 miles away, at 4 P.M. What was the average speed of the airplane? _____

UNIT 4 ■ TABLES, GRAPHS, AND AVERAGES

11. The local veterinary hospital has been very busy because of a recent cat virus. Monday they saw 18 cats, Tuesday 14 cats, Wednesday 15 cats, Thursday 13 cats, and Friday 15 cats. What was the average number of cats per day? _____

12. Alan raced his motorcycle on a mountain road, and traveled 45 miles in 90 minutes. How many miles per hour did he average? _____

5 MEASUREMENT

SKILLS YOU WILL NEED
    Basic math skills
    Careful reading

SKILLS YOU WILL LEARN
    Linear measurement
    Metric measurement
    Cooking measurement
    Temperature measurement

SITUATIONS IN WHICH YOU'LL USE THESE SKILLS
    At home
    At work

Measurement is a natural part of our life and everyday conversation: "How far did you drive?" "How cold is it going to be today?" "How tall are you?" "How much milk do we need from the store?" "I'll double the recipe for green beans, and make half the recipe for potato salad."

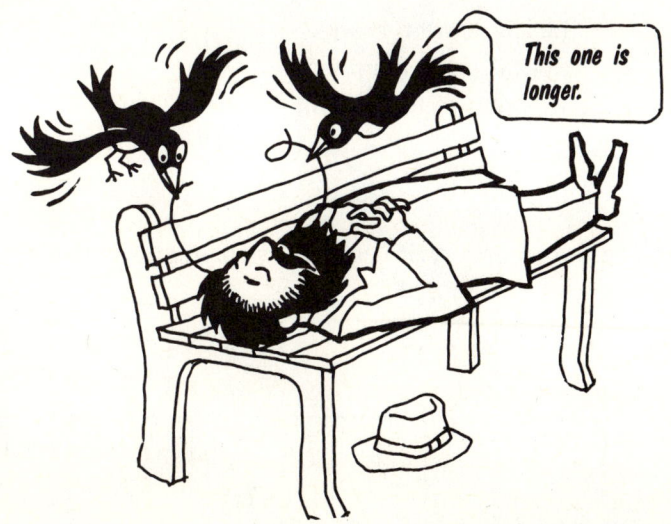

You will already be familiar with many of the measurement ideas in this unit. You have probably used them many times. But a review of measurement will help you become more accurate and precise in its use.

UNIT 5 ■ MEASUREMENT

LINEAR MEASUREMENT

Measuring with a ruler, yardstick, or tape measure is often necessary at home or on the job. Most rulers, yardsticks, and tape measures show inches and fractions of inches.

Some rulers are marked every $\frac{1}{4}$ inch.

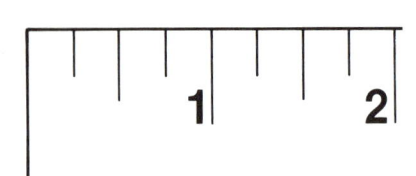

Some rulers are marked every $\frac{1}{8}$ inch.

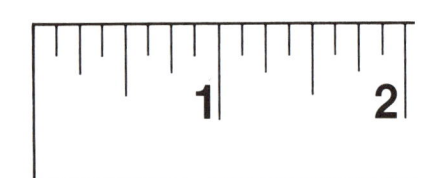

Some rulers are marked every $\frac{1}{16}$ inch.

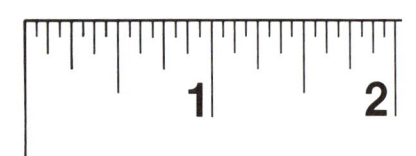

To read a measurement on a ruler, read the number of whole inches and the fractional part of an inch.

Example 1

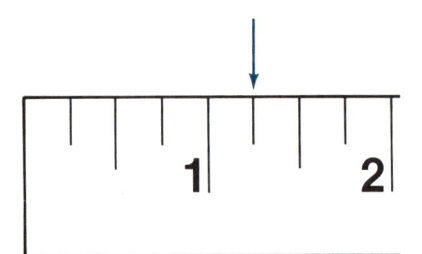

This arrow points to $1\frac{1}{4}$ inches.

Example 2

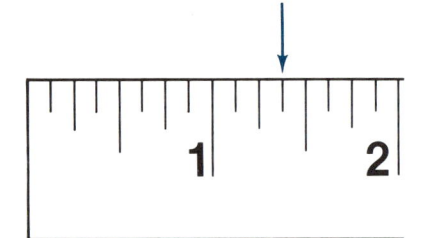

This arrow points to $1\frac{3}{8}$ inches.

Example 3

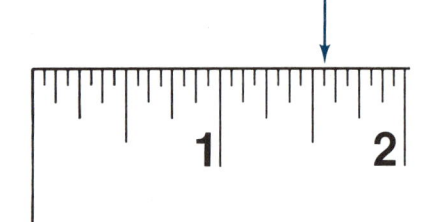

This arrow points to $1\frac{9}{16}$ inches.

> **HINT:** On most rulers and tape measures, the graduation marks for each different fractional size ($\frac{1}{2}$, $\frac{1}{4}$, $\frac{1}{8}$, $\frac{1}{16}$) are different lengths. The inch marks are longest. The $\frac{1}{2}$-inch, $\frac{1}{4}$-inch, $\frac{1}{8}$-inch, and $\frac{1}{16}$-inch marks decrease in length accordingly.

The arrows point to measurements on the rulers. Write each measurement on the line below the ruler.

1.

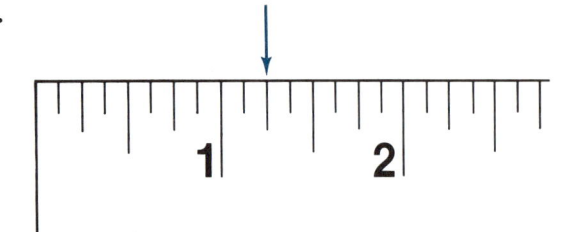

2.

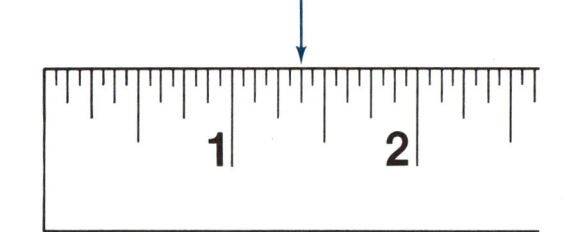

154

UNIT 5 ■ MEASUREMENT

3.

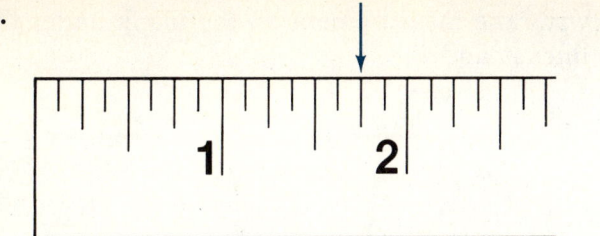

4.

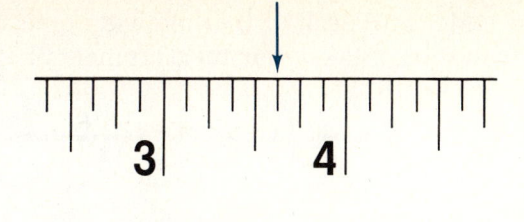

5.

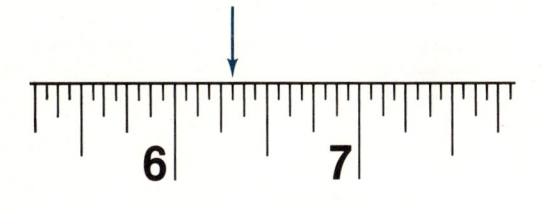

6.

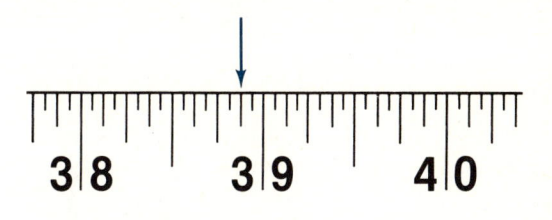

Measuring with a Ruler

Before you measure an object, look at your ruler or yardstick to make sure that a corner is not worn away. Using a worn ruler will make your measurement inaccurate. If you have a worn ruler, you can start the measurement at the 1-inch mark. If you do this, remember to subtract 1 from your measurement.

Example 1

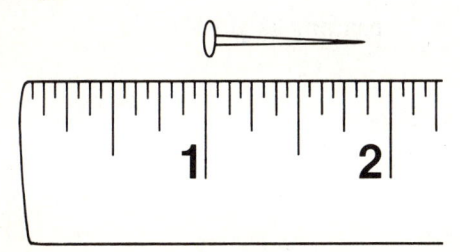

The tack measures $\frac{7}{8}$ inch, not $1\frac{7}{8}$ inches.

The measuring scale of some rulers does not begin at the end of the ruler. In this case, line the measuring scale with your object carefully.

Example 2

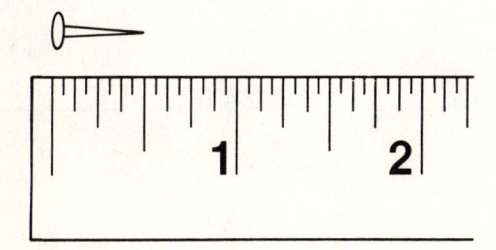

The tack measures $\frac{1}{2}$ inch.

UNIT 5 ■ MEASUREMENT

Measure each of the following objects and write the measurement on the blank under the illustration. Make your measurements between lines *A* and *B*.

1. 
2. 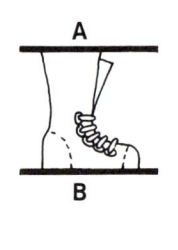
3. 
4.

When you read the fractional part of an inch on a ruler, use the reduced form of the fraction.

$\frac{2}{8}$ inch = $\frac{1}{4}$ inch $\frac{4}{8}$ inch = $\frac{1}{2}$ inch $\frac{6}{8}$ inch = $\frac{3}{4}$ inch

$\frac{4}{16}$ inch = $\frac{1}{4}$ inch $\frac{8}{16}$ inch = $\frac{1}{2}$ inch $\frac{12}{16}$ inch = $\frac{3}{4}$ inch

Example 3

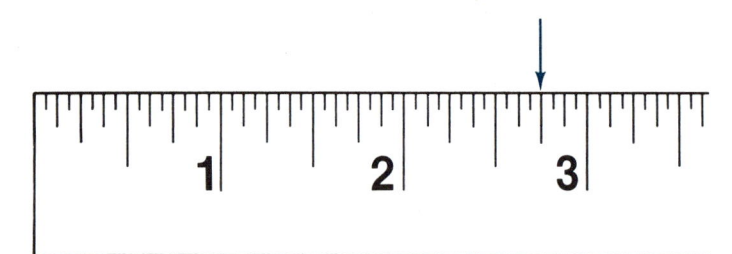

The arrow points to $2\frac{3}{4}$ inches. This is the same as $2\frac{6}{8}$ inches and the same as $2\frac{12}{16}$ inches. The first reading is preferred.

Answer the following questions.

1. How many $\frac{1}{8}$ are in each of the following?

   A. $\frac{1}{4}$ inch _____   B. 2 inches _____   C. $\frac{5}{8}$ inch _____

2. How many $\frac{1}{2}$ are in each of the following?

   A. 1 inch _____   B. $1\frac{1}{2}$ inches _____   C. 2 inches _____

3. How many $\frac{1}{4}$ are in each of the following?

   A. $\frac{1}{2}$ inch _____   B. 1 inch _____   C. $2\frac{1}{2}$ inches _____

4. How many $\frac{1}{16}$ are in each of the following?

   A. $\frac{3}{4}$ inch _____   B. $\frac{7}{8}$ inch _____   C. $1\frac{3}{8}$ inches _____

Draw an arrow to show the correct measurement on each partial ruler.

1. $\frac{5}{8}$ inch

2. $5\frac{1}{2}$ inches

3. $41\frac{7}{8}$ inches

4. $24\frac{13}{16}$ inches

5. $24\frac{3}{16}$ inches

6. $3\frac{3}{4}$ inches

Use a ruler to measure each of the lines below. Write the length above the line.

1. _____
2. _____
3. _____
4. _____
5. _____

6. _____
7. _____
8. _____
9. _____
10. _____

Use the illustration to answer the following questions.

1. What is the length of the guitar, between A and B?

2. What is the length of the center microphone, between C and D?

3. What is the length of the far right microphone, between E and F?

UNIT 5 ■ MEASUREMENT

Some of the graduation marks on the partial ruler are labeled with letters. Find out how far the letter *A* is from each of the following.

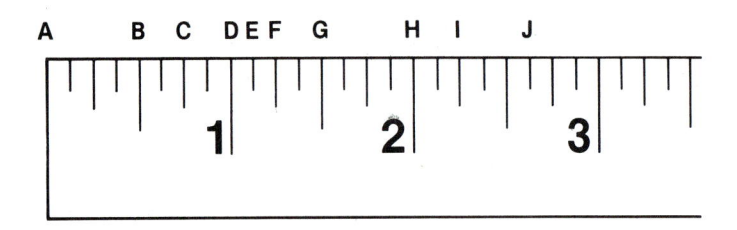

1. E _____
2. J _____
3. D _____
4. F _____
5. G _____
6. H _____
7. B _____
8. C _____

Equivalent Units

In addition to inches, you will often have to measure in feet and yards. Inches, feet, and yards are related as shown below.

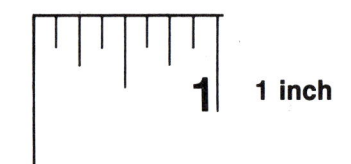

12 inches = 1 foot
36 inches = 1 yard
3 feet = 1 yard

Abbreviations

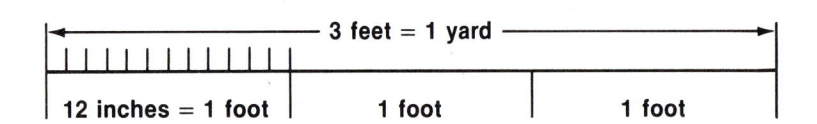

yard: yd
foot: ft or '
inch: in or "

The illustration above shows the relationship between inches, feet, and yards. It does *not* show actual size.

Write equivalent units for the following problems.

1. 2 feet = _____ inches
2. 1 yard = _____ inches
3. $3\frac{1}{2}$ feet = _____ inches
4. 2 yards = _____ feet
5. 6 yards = _____ feet
6. 6 feet = _____ yards
7. $2\frac{1}{2}$ feet = _____ inches
8. $1\frac{1}{2}$ yards = _____ feet
9. 2 yards = _____ inches
10. 18 inches = _____ feet

11. Allison purchased 5 yards of material. How many feet of fabric did she buy? _____

UNIT 5 ■ MEASUREMENT

Adding and Subtracting Measurements

To add or subtract with similar units of measurement, work as you would with fractions.

Example 1

$23\frac{3}{8}$ inches
$+6\frac{3}{8}$ inches
$29\frac{6}{8}$ inches = $29\frac{3}{4}$ inches

Example 2

$15\frac{3}{4}$ inches
$-7\frac{1}{4}$ inches
$8\frac{2}{4}$ inches = $8\frac{1}{2}$ inches

To add or subtract measurements that include inches, feet, and yards, be sure that you add or subtract inches and inches, feet and feet, and yards and yards. Change to a larger unit when possible.

Example 3

12 feet   3 inches
+ 4 feet   7 inches
16 feet 10 inches

Example 4

6 yards 2 feet
− 4 yards 1 foot
2 yards 1 foot

Example 5

13 feet 10 inches
+ 7 feet   4 inches
20 feet 14 inches = 21 feet 2 inches
    1 foot + 2 inches

Example 6

2 yards 1 foot
+ 1 yard   2 feet
3 yards 3 feet = 4 yards
         1 yard

Add and subtract in the following problems. Change to a larger unit when possible.

1. 4 feet   6 inches
   + 2 feet   9 inches

2. $3\frac{1}{2}$ yards
   $+2\frac{1}{4}$ yards

3. 5 feet   9 inches
   + 4 feet   3 inches

4. 40 yards 2 feet
   − 19 yards 1 foot

5. $\frac{5}{8}$ inch
   $-\frac{1}{8}$ inch

6. $9\frac{7}{8}$ inches
   $-2\frac{3}{8}$ inches

7. Walter purchased $2\frac{1}{8}$ yards of blue cotton, $1\frac{3}{4}$ yards of yellow wool, and $1\frac{5}{8}$ yards of white cotton. How many yards of fabric did he buy altogether?

UNIT 5 ■ MEASUREMENT

Solve the following linear measurement problems.

1. Use a ruler to measure the following lines. Write your answers on the blanks at the right.

   A. ─────────────

   B. ──────────────────

   C. ────────────────────────

2. Write equivalent units for each of the following.

   A. 8 inches 10 feet = _____ inches   D. 41 yards = _____ feet

   B. 11 yards 2 feet = _____ feet   E. 2 yards $2\frac{1}{2}$ feet = _____ inches

   C. 36 inches = _____ yards   F. 75 inches = _____ feet

3. Since there are 12 inches in 1 foot, how many feet are there in 576 inches?

4. Write 18 feet as yards.

5. Write 18 feet as inches.

6. Deena needed $2\frac{5}{8}$ yards of material for pants and $1\frac{7}{8}$ yards for a blouse. What was the total amount of yardage she needed?

UNIT 5 ■ MEASUREMENT

7. On a map, 3 inches represents 135 miles. How many miles are represented by 1 inch? _____

8. Ernie needed 99 inches of window trim. How many feet of trim did he need? _____

9. Fran's skirt length is 22 inches. Her pattern measures $25\frac{1}{2}$ inches. How much does she need to turn up for a hem? _____

10. Miguel is 74 inches tall. How tall is he in feet and inches? _____

11. Fred needed to purchase lumber for a fence. He needed a board 18 feet 6 inches long and a board 6 feet 8 inches long. How much lumber did he need to buy? _____

UNIT 5 ■ MEASUREMENT

12. A map drawing represents 50 miles as 1 inch. If the drawn distance between two cities is $6\frac{1}{2}$ inches, how far apart are the cities?

13. The scale for a building blueprint shows every 10 feet as 3 inches. If the building is 75 feet wide, how wide will the building be on the blueprint?

14. A sheet of plywood measures $4\frac{1}{2}$ feet wide. The plywood is to be cut exactly in half. How wide will each piece be if the saw blade removes $\frac{1}{8}$ inch?

15. An enlarged photograph is proportional to the size of its negative. If the negative measures 1.5 inches wide by 2.5 inches long, how long will a print be that measures 10 inches wide?

METRIC MEASUREMENT

The United States system of measurement was developed gradually in England over hundreds of years. Originally, many units of measurement in the U.S. system were based on parts of a person's body. For example, a foot was the length of a person's foot. Although the units of measurement in the U.S. system are now standardized, the relationship between equivalent units does not follow a logical pattern: There are 12 inches in a foot, 16 ounces in a pound, 4 quarts in a gallon, 2,000 pounds in a ton, 1,760 yards in a mile. Because you are familiar with the U.S. system, it seems fairly easy for you to use it.

Most of the world uses the *metric* system of measurement. The metric system is a decimal system and is based on multiples of 10. The metric system is really much easier to use than the U.S. system. The metric system has three basic units of measurement.

Meter A meter is used to measure how long, how wide, or how high something is.

Gram A gram is used to measure how heavy something is.

Liter A liter is used to measure how much is needed to fill something.

Decide which unit of measurement would be used to measure the following. Write the word *meter*, *gram*, or *liter*.

1. Your weight _____
2. Your height _____
3. A glass of water _____
4. A picture frame _____
5. A pork roast _____
6. A tank of gasoline _____

Practice using the meter, gram, and liter by deciding on the correct unit of measurement. Put an X in the appropriate column.

		meter	gram	liter
1.	Desk (width)			
2.	Coffee cup (volume)			
3.	Window (length)			
4.	Piece of chalk (weight)			
5.	Chalkboard (width)			
6.	Glass (volume)			
7.	Pencil (length)			
8.	Pencil (weight)			
9.	Milk carton (volume)			

UNIT 5 ■ MEASUREMENT

Comparing Metric and U.S. Measurements

It is important to understand the metric system in terms of everyday objects. Practice with metric units makes the metric system easy to use.

Length is measured in *meters* (m).
A meter is a little longer than a yard (about $3\frac{3}{4}$ inches longer).

```
|            1 meter            |
```

```
|            1 yard           |
```

Weight is measured in *grams* (g).
A paper clip weighs about 1 gram.   A nickel weighs about 5 grams.

Volume is measured in *liters* (l).
A liter is a little larger than a quart.

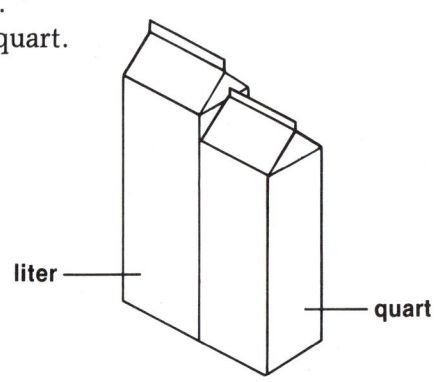

liter —   — quart

Answer the following questions. For Problems 2-5, write the letter of the correct answer on the blank at the right.

1. Which is longer, a yard or a meter? _____

2. A liter tells
   A. how long   C. how much   E. how cold
   B. what time   D. how far   _____

3. What metric unit would be used to measure the weight of a tent?
   A. gram   B. meter   C. liter   _____

4. About how often would a growing teenager drink 1 liter of milk?
   A. each week   B. each day   C. each year   _____

5. Weight is measured in
   A. grams   B. meters   C. liters   _____

164   UNIT 5 ■ MEASUREMENT

Metric Prefixes

The metric system uses prefixes to show measurements larger or smaller than a meter, gram, or liter.

If something is *much larger* than a meter, gram, or liter, the prefix *kilo* is added. This makes the unit of measurement 1,000 times larger.

1 kilometer = 1,000 meters (about 0.6 mile)
1 kilogram = 1,000 grams (about 2.2 pounds)

If something is *much smaller* than a meter, gram, or liter, the prefix *centi* is added. This makes the unit of measurement $\frac{1}{100}$ as large. This is like our money system in which a cent is $\frac{1}{100}$ of a dollar.

1 centimeter = $\frac{1}{100}$ meter
1 centigram = $\frac{1}{100}$ gram
1 centiliter = $\frac{1}{100}$ liter

1 inch is about 2.5 centimeters

3 centimeters

If something is even smaller, the prefix *milli* is added. This makes the unit of measurement $\frac{1}{1,000}$ as large.

1 millimeter = $\frac{1}{1,000}$ meter
1 milligram = $\frac{1}{1,000}$ gram
1 milliliter = $\frac{1}{1,000}$ liter

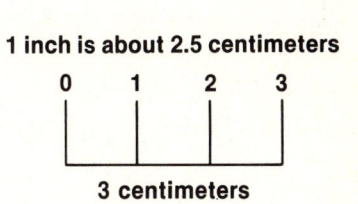

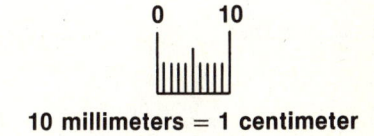

10 millimeters = 1 centimeter

Memorize these prefixes:

Prefix

kilo kilo means 1,000.
  1 kilometer (km) is 1,000 meters.
  1 kilogram (kg) is 1,000 grams.

centi centi means $\frac{1}{100}$, or 0.01.
  1 centimeter (cm) is $\frac{1}{100}$ of a meter.
  1 centiliter (cl) is $\frac{1}{100}$ of a liter.

milli milli means $\frac{1}{1,000}$, or 0.001.
  1 millimeter (mm) is $\frac{1}{1,000}$ of a meter.
  1 milligram (mg) is $\frac{1}{1,000}$ of a gram.

Write *kilo*, *centi*, or *milli* to complete each word.

1. $\frac{1}{100}$ meter is a _____ meter.

2. 1,000 meters is a _____ meter.

3. $\frac{1}{1,000}$ gram is a _____ gram.

4. 1,000 grams is a _____ gram.

5. $\frac{1}{1,000}$ liter is a _____ liter.

UNIT 5 ■ MEASUREMENT

Write $\frac{1}{1,000}$, $\frac{1}{100}$, or 1,000 for the following prefixes.

1. kilo _____

2. centi _____

3. milli _____

Circle the *larger* unit of measurement.

1. milligram   kilogram   5. kilogram   gram

2. kilometer   centimeter   6. liter   milliliter

3. centiliter   milliliter   7. kilometer   meter

4. meter   centimeter

Organize the following terms from smallest to largest in the table below.
liter, milligram, kilometer, centimeter, milliliter, meter, kilogram, millimeter, centigram, centiliter, gram

Volume	Length/Width	Weight
_____	_____	_____
_____	_____	_____
_____	_____	_____
_____	_____	_____

Metric Units Written as Decimals

Metric measurements *less than 1* can be written in decimal form because the metric system is based on multiples of 10.

Example 1   3 centimeters = 0.03 meter (3 hundredths of a meter)
3 millimeters = 0.003 meter (3 thousandths of a meter)

15 centigrams = 0.15 gram (15 hundredths of a gram)
15 milligrams = 0.015 gram (15 thousandths of a gram)

57 centiliters = 0.57 liter (57 hundredths of a liter)
57 milliliters = 0.057 liter (57 thousandths of a liter)

Write the following metric measurements in decimal form.

1. 1 centimeter = __0.01__ meter
2. 18 centimeters = _____ meter
3. 35 centimeters = _____ meter
4. 98 milligrams = _____ gram
5. 8 centiliters = _____ liter
6. 6 millimeters = __0.006__ meter
7. 23 millimeters = _____ meter
8. 118 milligrams = _____ gram
9. 15 milliliters = _____ liter
10. 325 millimeters = _____ meter

Metric measurements *greater than 1* can be written in two ways.

Example 2

Whole number and decimal:

    1.5 meters
    1.5 liters
    1.5 grams

Centimeters, centiliters, or centigrams:

    150 centimeters
    150 centiliters
    150 centigrams

Write the following metric measurements in another way.

1. 3 meters = __300 centimeters__
2. 5 meters = _____
3. 2.5 liters = _____
4. 20 grams = _____
5. 8.75 meters = _____
6. 3.75 grams = _____
7. __6 meters__ = 600 centimeters
8. _____ = 225 centigrams
9. _____ = 1,500 centimeters
10. _____ = 55 centiliters
11. _____ = 450 centigrams
12. _____ = 405 centimeters

Measurement Using Meters

A meter is separated into 100 equal parts. Each part is a centimeter. Each centimeter is also separated into 10 equal parts. Each of these parts is a millimeter. (Remember, a meter is a little longer than a yard.)

A meter stick has 100 centimeter marks and 1,000 millimeter marks.

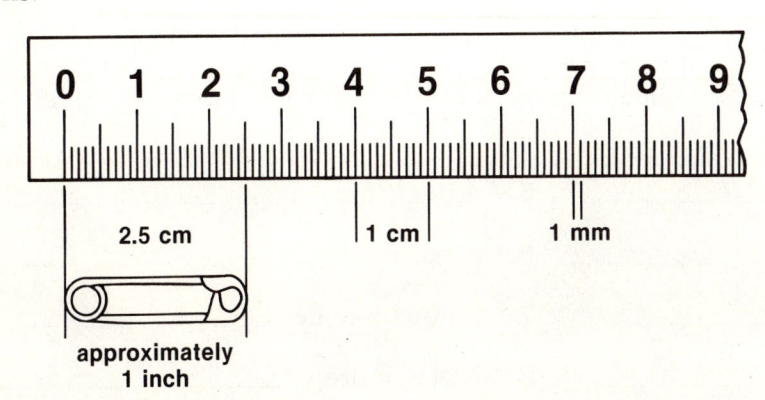

UNIT 5 ■ MEASUREMENT

Answer the following questions.

1. A 1-inch safety pin is about how many centimeters long? _____
2. A 1-inch safety pin is about how many millimeters long? _____
3. How many centimeters equal 1 meter? _____
4. How many millimeters equal 1 centimeter? _____
5. This line is 4 centimeters long: ———
   How many millimeters long is it? _____
6. This line is 50 millimeters long: ———
   How many centimeters long is it? _____
7. The Mississippi River is longer than a kilometer. Four laps around a football field is also longer than a kilometer. Write 2 more examples that are longer than a kilometer.

_____

_____

8. Write *meters*, *centimeters*, or *kilometers* for each of the following.
   A. A city building could be 100 __?__ high. _____
   B. I could walk 2 __?__ to a party. _____
   C. My pen is about 15 __?__ long. _____

Measurement Using Grams

A kilogram is about 2.2 pounds. There are 1,000 grams in a kilogram. Recall that a paper clip weighs about 1 gram.

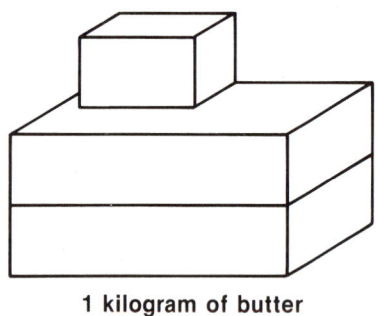

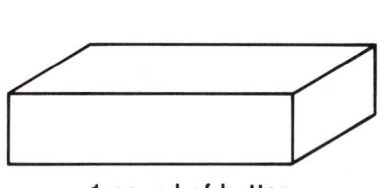

1 kilogram of butter   1 pound of butter   1 gram

Would the following objects be weighed in grams or kilograms? Write *grams* or *kilograms* after each word.

1. A flower _____   5. A dog _____
2. A table _____   6. A person _____
3. A cupcake _____   7. An automobile _____
4. A pencil _____   8. A cough drop _____

For the following problems, write the letter of the best answer in the blank at the right.

1. Weight is measured in
   A. grams   B. meters   C. liters

2. 1,000 grams is
   A. 1 kilogram   B. 1 kilometer   C. 1 milligram

3. A paper clip weighs about
   A. 1 kilogram   B. 1 centigram   C. 1 gram

4. Neil Taylor, champion swimmer, weighs about
   A. 30 kilograms   B. 50 kilograms   C. 70 kilograms

5. Five nickels weigh about
   A. 25 grams   B. 50 grams   C. 1 kilogram

6. One kilogram equals 1,000 grams; 500 grams equals
   A. $\frac{1}{2}$ kilogram   B. 2 kilograms   C. 2 pounds

7. A weight lifter can lift 275,000 grams. How many kilograms is this?
   A. 27.5 kilograms   B. 275 kilograms   C. 2,750 kilograms

Use your knowledge of metric measurement to answer the following questions.

1. If a normal walking step is 40 centimeters, how many steps will it take to walk 1 kilometer?

2. A ceramics student made a pitcher that holds
   A. 2 liters
   B. 10 liters
   C. 25 liters

3. If you can drive 128 kilometers on 16 liters of gas, how far can you drive on 1 liter?

UNIT 5 ■ MEASUREMENT

4. A board 117 centimeters long is cut into 2 pieces so that 1 piece is 2 times as long as the other. How long are the pieces? (Disregard waste from the cuts.)

_____

5. Ruth bought 45.6 liters of gasoline. If 3.8 liters equals 1 gallon, how many gallons of gas did she buy?

_____

6. A coffee cup holds
    A. less than a liter
    B. a liter
    C. more than a liter

_____

7. Virginia Austin's family of four enjoys ice cream. Each week the family eats about
    A. 1 liter
    B. 10 liters
    C. 100 liters

_____

8. On a map on which every centimeter represents 2 miles, Los Angeles and San Francisco are 200 centimeters apart. What is the approximate mile distance between the two cities?

_____

9. One pill contains 6 milligrams. How many milligrams are in a bottle of 36 pills?

_____

10. For the pills in Problem 9, how many pills can be made from 3 grams of the drug?

_____

COOKING MEASUREMENT

Knowing basic cooking measurements and equivalents makes it easier to cook and follow recipes. Study the following table of measurement equivalents.

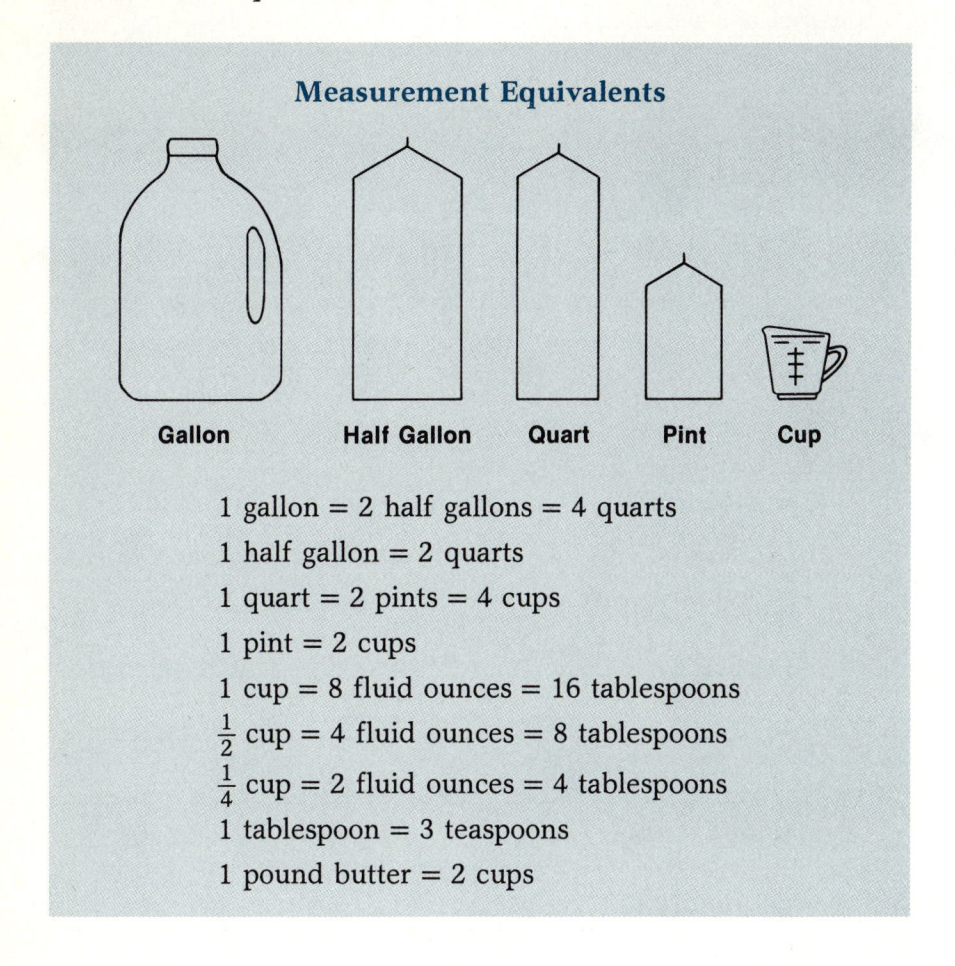

Measurement Equivalents

Gallon   Half Gallon   Quart   Pint   Cup

1 gallon = 2 half gallons = 4 quarts
1 half gallon = 2 quarts
1 quart = 2 pints = 4 cups
1 pint = 2 cups
1 cup = 8 fluid ounces = 16 tablespoons
$\frac{1}{2}$ cup = 4 fluid ounces = 8 tablespoons
$\frac{1}{4}$ cup = 2 fluid ounces = 4 tablespoons
1 tablespoon = 3 teaspoons
1 pound butter = 2 cups

Using the table above, write the equivalent units for the following.

1. 8 tablespoons = _____ cup
2. 4 cups = _____ pints
3. 4 quarts = _____ gallon
4. 12 fluid ounces = _____ cups
5. 4 tablespoons = _____ cup
6. 3 tablespoons = _____ teaspoons
7. $1\frac{1}{2}$ quarts = _____ cups
8. $\frac{1}{2}$ cup = _____ fluid ounces
9. $\frac{3}{4}$ cup = _____ tablespoons
10. 1 cup butter = _____ pound
11. 6 pints = _____ quarts
12. 16 tablespoons = _____ cup
13. 12 tablespoons = _____ cup
14. 2 quarts = _____ gallon
15. 1 gallon = _____ pints
16. $\frac{1}{2}$ gallon = _____ quarts
17. $\frac{1}{4}$ cup = _____ tablespoons
18. 2 cups = _____ fluid ounces

UNIT 5 ■ MEASUREMENT

Measuring Equipment

Common kitchen equipment used to measure various kinds of ingredients are shown below.

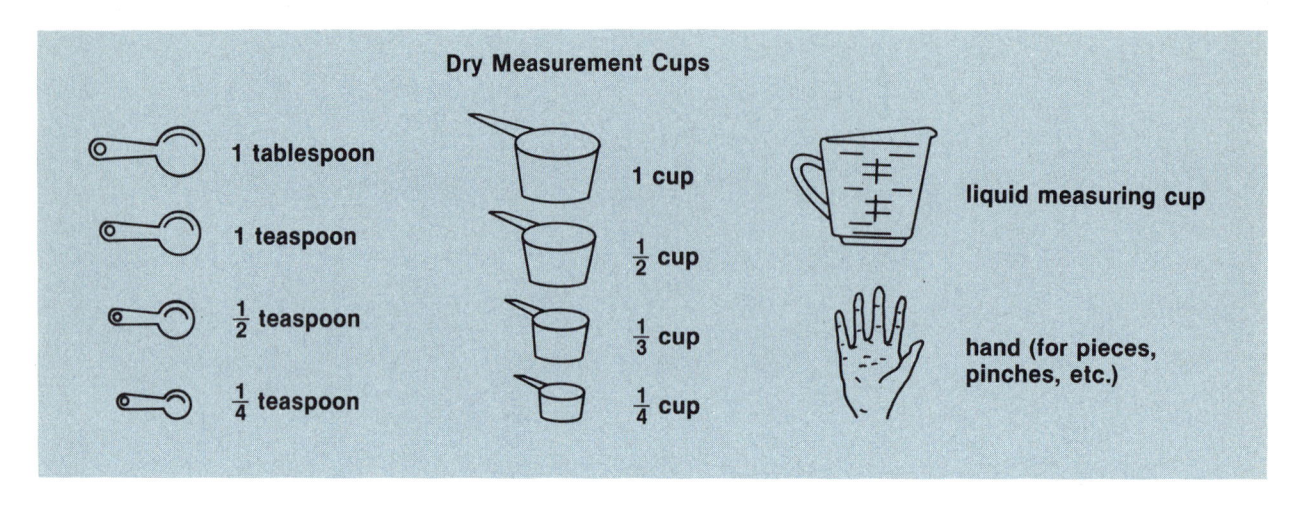

Dry Measurement Cups

- 1 tablespoon
- 1 teaspoon
- $\frac{1}{2}$ teaspoon
- $\frac{1}{4}$ teaspoon
- 1 cup
- $\frac{1}{2}$ cup
- $\frac{1}{3}$ cup
- $\frac{1}{4}$ cup
- liquid measuring cup
- hand (for pieces, pinches, etc.)

Write the equipment you would use to measure each of the following. If you would have to use the measuring equipment more than once, write the number of times you would use it.

		Equipment	Times Used
1.	2 cups flour	1 cup	2
2.	$\frac{2}{3}$ cup sugar		
3.	2 teaspoons salt		
4.	$\frac{1}{4}$ cup milk		
5.	$\frac{3}{4}$ cup oatmeal		
6.	dash of nutmeg		
7.	$\frac{1}{2}$ cup celery		
8.	2 tablespoons water		
9.	1 cup shortening		
10.	$\frac{3}{4}$ teaspoon baking powder		
11.	$\frac{2}{3}$ cup ginger ale		
12.	1 cup relish		
13.	pinch of pepper		
14.	$\frac{1}{6}$ cup brown sugar		
15.	$\frac{1}{2}$ tablespoon heavy cream		
16.	$\frac{1}{4}$ pound margarine		

Increasing and Decreasing Recipes

Recipes are not always written in the amounts you need. Many times you will have to reduce or increase a recipe. To cut a recipe in half, *divide* all the amounts by 2. To double a recipe, *multiply* all the amounts by 2.

Halve and double the following two recipes. Be sure to put the new amounts in the form that is easiest to measure.

Chocolate Nut Cake

Recipe	Half	Double
$\frac{1}{2}$ cup butter		
1 cup sugar		
2 eggs		
2 ounces chocolate		
$\frac{1}{3}$ cup buttermilk		
1 teaspoon vanilla		
$1\frac{3}{4}$ cup flour		
$\frac{1}{2}$ teaspoon soda		
$\frac{1}{2}$ teaspoon salt		
$\frac{1}{4}$ cup chopped nuts		

Fluffy Dessert

Recipe	Half	Double
$\frac{2}{3}$ cup sugar		
1 tablespoon gelatin		
1 cup crushed fruit		
2 eggs		
$\frac{1}{4}$ teaspoon cream of tartar		
$\frac{1}{3}$ cup sugar		
$\frac{1}{2}$ cup whipped cream		

UNIT 5 ■ MEASUREMENT

Using Cooking Equivalents

When cooking it is helpful to know food equivalents. Food equivalents are the same food measured two different ways, such as cooked and uncooked or pounds and cups. Many cookbooks have food equivalent tables similar to the one below.

Common Food Equivalents

Food	Amount	Approximate Measure
Cheddar cheese	1 pound	4 cups, grated
Chocolate chips	6-ounce package	1 cup
Cream, whipping	1 cup, liquid	2 cups whipped
Flour	1 pound	$3\frac{1}{2}$ cups
Lemon	1 medium	$2\frac{1}{2}$ tablespoons juice
Macaroni	7 ounces, uncooked	4 cups, cooked
Rice	1 cup, uncooked	3 cups, cooked
Sugar, brown	1-pound box	$2\frac{1}{3}$ cups
Sugar, powdered	1-pound box	4 cups
Walnuts	1 pound, shelled	4 cups

Use the food equivalents table to find the following.

1. 6 cups whipped cream = _____ cups liquid cream
2. 5 tablespoons lemon juice = _____ lemons
3. 2 pounds brown sugar = _____ cups
4. 6 cups powdered sugar = _____ pounds
5. 1 cup grated cheddar cheese = _____ pounds
    = _____ ounces
6. 7 cups flour = _____ pounds
7. $2\frac{1}{2}$ cups uncooked rice = _____ cooked rice
8. 3 cups walnuts = _____ pounds = _____ ounces
9. 8 ounces chocolate chips = _____ cups
10. 6 cups cooked macaroni = _____ ounces uncooked macaroni
11. To make peach ice cream, you need $\frac{3}{4}$ cup of cream for every 3 quarts of crushed peaches. How much cream is needed for 12 quarts of peaches? _____

174 UNIT 5 ■ MEASUREMENT

TEMPERATURE MEASUREMENT

Temperature is measured using a *thermometer*. You have probably used two different types of thermometers. One type of thermometer is used to measure air temperature. A different type of thermometer is used to measure body temperature.

Most thermometers have a thin glass tube with liquid inside, usually mercury or alcohol. The liquid expands or contracts as the air or body temperature changes. To read a thermometer, see where the expanded liquid meets the scale on the thermometer.

Example 1   A thermometer that measures air temperature may have a Fahrenheit scale or a Celsius scale. On both scales, each mark represents 2 degrees.

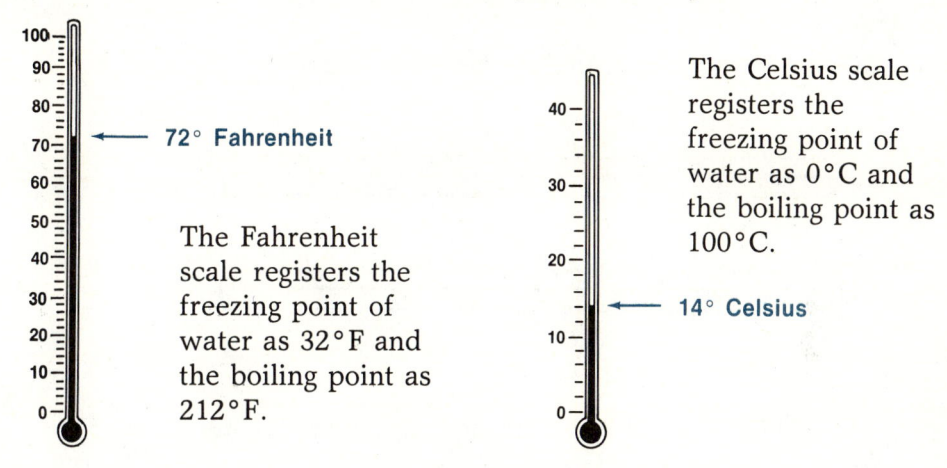

← 72° Fahrenheit

The Fahrenheit scale registers the freezing point of water as 32°F and the boiling point as 212°F.

The Celsius scale registers the freezing point of water as 0°C and the boiling point as 100°C.

← 14° Celsius

Answer the following questions by reading the three thermometers.

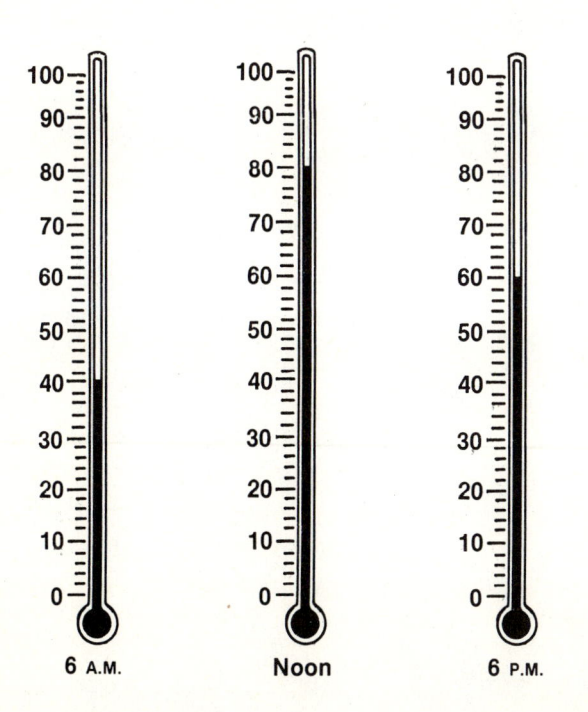

6 A.M.   Noon   6 P.M.

1. What was the temperature at 6 A.M.?

    _____

2. What was the difference between the highest and lowest temperatures?

    _____

3. What was the average temperature?

    _____

4. What was the difference in temperature between noon and 6 P.M.?

    _____

UNIT 5 ■ MEASUREMENT

Example 2 Thermometers that measure body temperature use the Fahrenheit scale. Each small mark represents 0.2°. Normal mouth temperature is 98.6°, which is marked with an arrow on the thermometer.

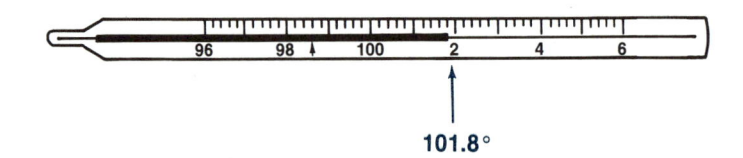

101.8°

Read the temperature on each of the following thermometers and record it on the line to the right.

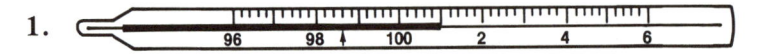

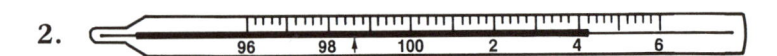

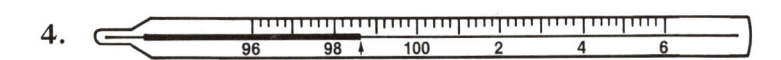

1. _____

2. _____

3. _____

4. _____

5. To find how much above normal a temperature is, subtract 98.6° from the thermometer reading. Use your answers to Problems 1-4 to answer the following questions.

   A. How much above normal is thermometer 1?   _____

   B. How much above normal is thermometer 2?   _____

   C. How much above normal is thermometer 3?   _____

   D. How much above normal is thermometer 4?   _____

Use the thermometers below to answer the questions.

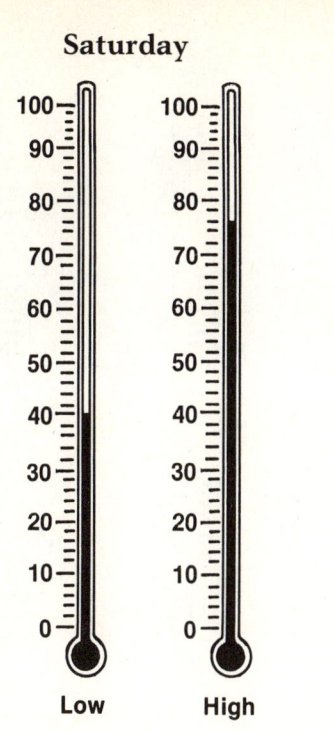

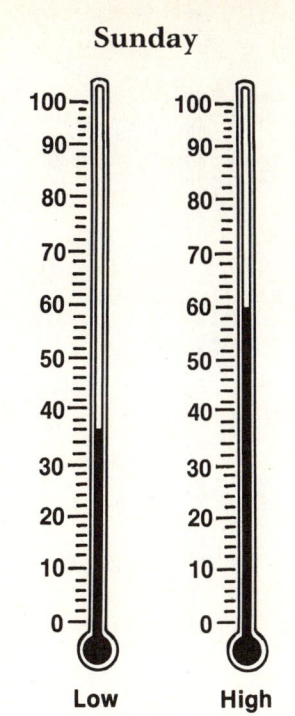

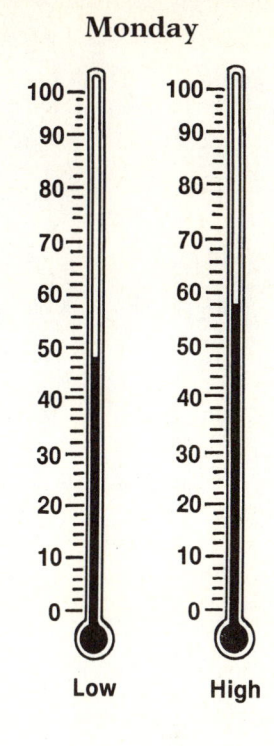

| Saturday | Sunday | Monday |
| Low High | Low High | Low High |

1. **A.** Which day had the lowest temperature? _____

   B. What was that temperature? _____

2. **A.** Which day had the highest temperature? _____

   B. What was that temperature? _____

3. What was the difference between the high and low temperatures on Monday? _____

4. What was the average low temperature for the 3 days? (Round your answer to the nearest degree.) _____

5. What was the average high temperature for the 3 days? (Round your answer to the nearest degree.) _____

UNIT 5 ■ MEASUREMENT

Fill in the thermometers as indicated in the following problems.

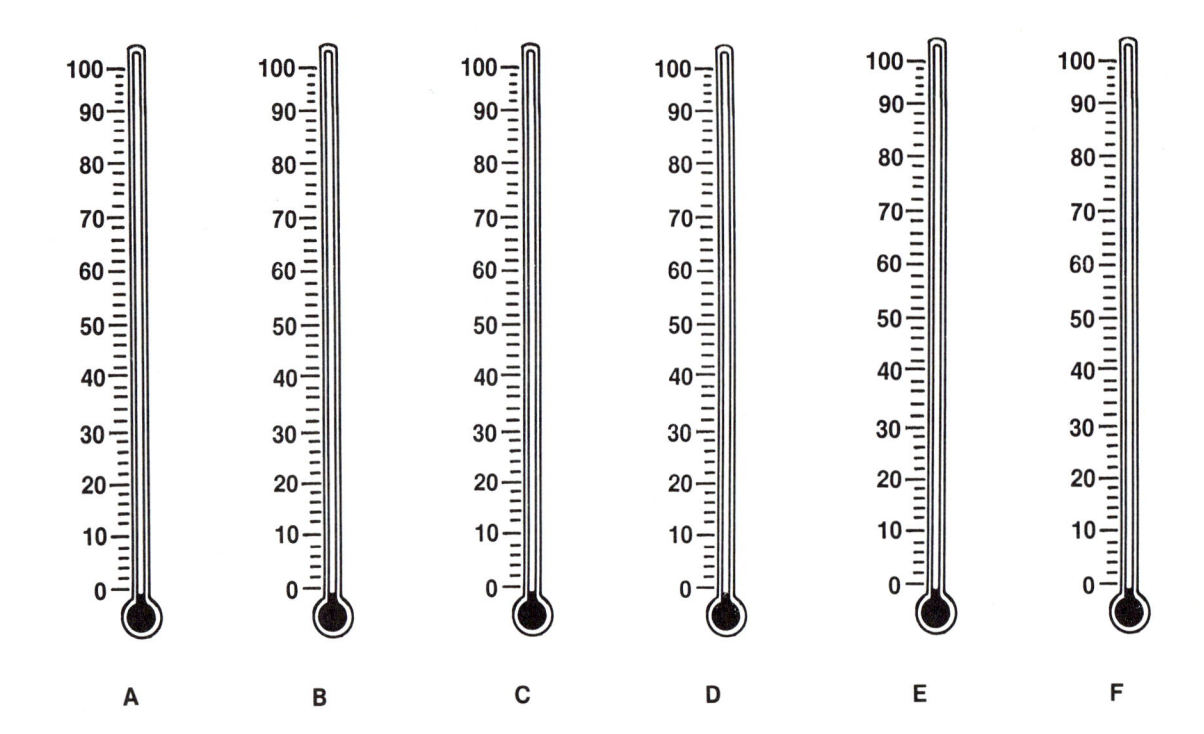

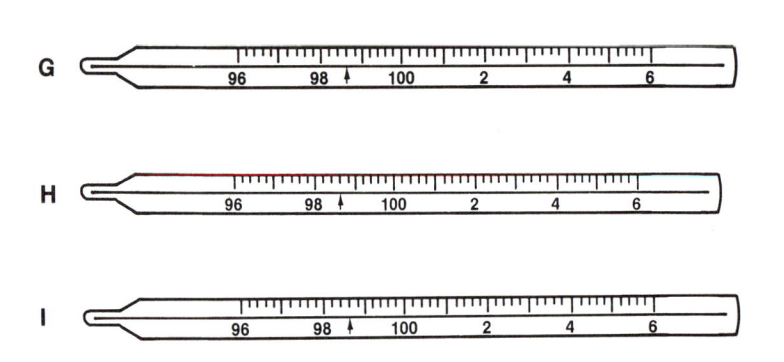

1. Fill in thermometer A to read 52°.
2. Fill in thermometer B to read 44°.
3. Fill in thermometer C to read 88°.
4. Fill in thermometer D to read 96°.
5. Fill in thermometer E to read 18°.
6. Fill in thermometer F to be twice as high as thermometer E.
7. Fill in thermometer G to read 103.2°.
8. Fill in thermometer H to read a normal temperature.
9. Fill in thermometer I to read 100.8°.

UNIT 5 ■ MEASUREMENT

6 BANKING

SKILLS YOU WILL NEED
  Basic math skills

SKILLS YOU WILL LEARN
  Opening a bank account
  Endorsing a check
  Making a deposit
  Writing checks
  Filling out a check register
  Identifying parts of a bank statement
  Using an automatic teller machine (ATM)
  Reconciling a bank statement

SITUATIONS IN WHICH YOU'LL USE THESE SKILLS
  Shopping
  Paying bills
  Keeping a record of money spent

Banks offer savings accounts, checking accounts, safe deposit boxes, financial advice, and a variety of loan services. Savings accounts and checking accounts are safe places to keep money. Carrying a checkbook is safer than carrying large amounts of money when you shop. Also, a monthly checking account statement from the bank provides proof that a bill has been paid and helps you keep track of your money. A safe deposit box is a box located in the bank's vault where important papers and other valuables can be kept.

You will probably not need many of the services offered by banks just yet. You may not have enough money to invest to take advantage of a bank's financial advice. But savings and checking accounts make sense for almost everyone.

Banks and accounts within banks vary in the amount of interest paid and the amount charged for services. Be sure to check with your own bank for variations.

UNIT 6 ■ BANKING

OPENING A BANK ACCOUNT

Throughout your life you will probably use many different banking services. But your first dealings with a bank will most often be in opening a savings or checking account.

A *savings account* is not only a safe place to keep your money, it also earns you money while you save. A bank pays you interest on the money in your savings account.

A *checking account* provides a convenient way to shop and pay bills. Checks are safe to send in the mail. A checking account also provides a record of where money has been spent.

When opening either a savings or checking account, you need to fill out a *signature card*. The signature card must be signed the way you plan to sign withdrawal slips or checks. If you wish to have someone else in the family use your account, that person would also have to sign the signature card.

Fill out the signature card below for yourself. Assume you will be the only individual to sign on a regular checking account.

DIABLO BANK

ACCOUNT AUTHORIZATION

A. Type of Account(s)
- ☐ Regular Checking
- ☐ Checking/Interest Plan
- ☐ Market Rate Checking
- ☐ Savings
- ☐ Young Saver
- ☐ Time Deposit

B. Account Ownership   Please Check One
- ☐ Individual
- ☐ Community Property
- ☐ Trust For Named Beneficiary
- ☐ Tenants in Common
- ☐ Joint Tenants

If Trust, Give Name of Beneficiary(ies) _____

ACCOUNT NAME _____

Number of Signatures Required to Withdraw Cash

X _____   X _____
  Customer 1 Signature   Customer 2 Signature

C. Customer Information

Please Print	Customer 1	Customer 2
Social Security No.		
Street Address, Apt. #		
City		
State, Zip Code		
Phone: Home	( )	( )
Business	( )	( )
Occupation		
Employer		
Birth Date		
Birthplace		
Mother's Maiden Name		

DEPOSIT SLIPS AND CHECKS

Savings and checking accounts have several different kinds of forms. It is important that these forms be filled out correctly. You don't want the bank to make a mistake with your money because you made a mistake.

Endorsing a Check

Before you can cash a check or deposit it into your account, you must *endorse* the check. To endorse a check, you sign your name on the back of the check the same as it is written on the front of the check.

When you cash a check, a bank will usually want you to endorse the check while the teller watches you. If the teller suspects a person is forging a signature, he or she will check the signature card for the account.

Example 1 Endorsing a check to cash it.

Example 2 Endorsing a check to deposit it to an account.

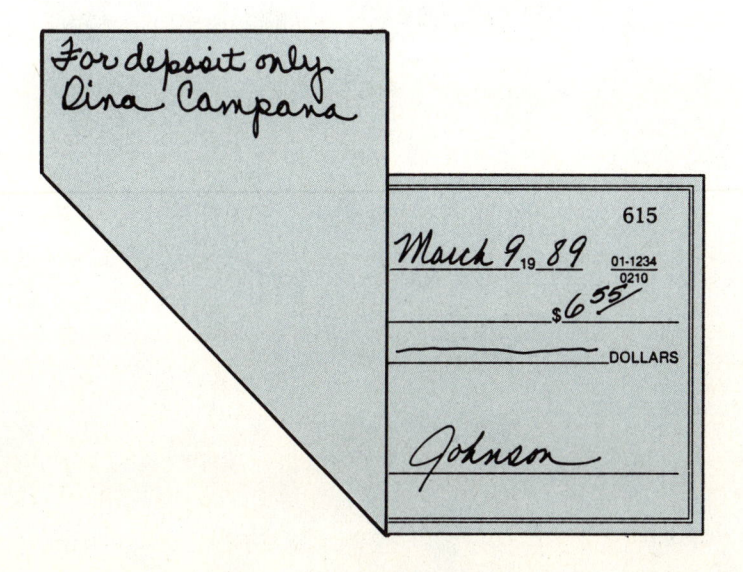

UNIT 6 ■ BANKING

Making a Deposit

After filling out and signing a signature card, you can deposit money or checks to your new account. If the new account is a checking account, you must make a deposit before you can write a check.

The steps below explain how to fill out a deposit slip. Study the deposit slip in the example as you read the steps.

1. Write the amount of paper money on the line marked *currency*.
2. Write the amount of change on the line marked *coin*.
3. List the amount of each check on the lines marked *checks*. If you have more checks than spaces on the front of the deposit slip, list additional checks on the back of the deposit slip. Endorse all checks written to you.
4. Add currency, coins, and checks. Record this amount on the line marked *subtotal*.
5. If you wish to receive cash, write the amount you want on the *less cash received* line.
6. Subtract the amount of cash you wish to receive.
7. Write your deposit on the line marked *total*.
8. Write your name and address and the date and sign your name if you received cash.

Example

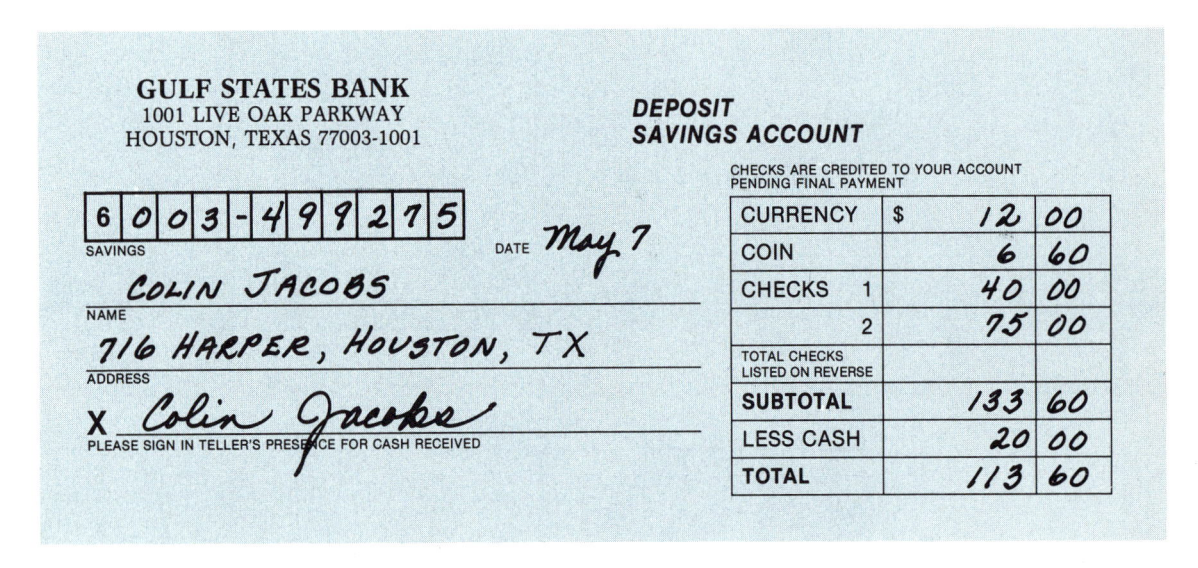

	GULF STATES BANK 1001 LIVE OAK PARKWAY HOUSTON, TEXAS 77003-1001		DEPOSIT SAVINGS ACCOUNT		
			CHECKS ARE CREDITED TO YOUR ACCOUNT PENDING FINAL PAYMENT		
6003-499275 SAVINGS		DATE May 7	CURRENCY	$ 12	00
			COIN	6	60
COLIN JACOBS NAME			CHECKS 1	40	00
			2	75	00
716 HARPER, HOUSTON, TX ADDRESS			TOTAL CHECKS LISTED ON REVERSE		
X Colin Jacobs PLEASE SIGN IN TELLER'S PRESENCE FOR CASH RECEIVED			SUBTOTAL	133	60
			LESS CASH	20	00
			TOTAL	113	60

Answer the following questions by looking at the example.

1. How much paper money and change was deposited? _____

2. How much cash was received? _____

3. Who made the deposit? _____

UNIT 6 ■ BANKING

Fill out the following deposit slips with the information given.

1. On April 5, deposit $3 in cash, 50¢ in coins, a check for $150, and a check for $35. You would like $50 in cash.

GULF STATES BANK
1001 LIVE OAK PARKWAY
HOUSTON, TEXAS 77003-1001

DEPOSIT
SAVINGS ACCOUNT

CHECKS ARE CREDITED TO YOUR ACCOUNT PENDING FINAL PAYMENT

CURRENCY	$	
COIN		
CHECKS 1		
2		
TOTAL CHECKS LISTED ON REVERSE		
SUBTOTAL		
LESS CASH		
TOTAL		

6 ☐☐☐☐☐☐☐
SAVINGS   DATE

NAME _____

ADDRESS _____

X _____
PLEASE SIGN IN TELLER'S PRESENCE FOR CASH RECEIVED

2. On January 16, deposit $33.50 in cash ($32 in paper, $1.50 in coins) and checks for $46.74, $52.80, and $15.40.

DIABLO BANK
2730 SALVIO STREET
CONCORD, CALIFORNIA 94519

CHECKING ACCOUNT DEPOSIT TICKET

CASH	CURRENCY		
	COIN		
LIST CHECKS SINGLY			
TOTAL FROM OTHER SIDE			
TOTAL			
LESS CASH RECEIVED			
NET DEPOSIT			

USE OTHER SIDE FOR ADDITIONAL LISTING

BE SURE EACH ITEM IS PROPERLY ENDORSED

DATE _____ 19 ___

_____
SIGN HERE FOR LESS CASH IN TELLER'S PRESENCE

⑇021012348⑇0340⑈04370015 77⑈

3. On February 2, deposit checks for $25.50 and $93.40. Receive cash of $15.

GULF STATES BANK
1001 LIVE OAK PARKWAY
HOUSTON, TEXAS 77003-1001

DEPOSIT
SAVINGS ACCOUNT

CHECKS ARE CREDITED TO YOUR ACCOUNT PENDING FINAL PAYMENT

CURRENCY	$	
COIN		
CHECKS 1		
2		
TOTAL CHECKS LISTED ON REVERSE		
SUBTOTAL		
LESS CASH		
TOTAL		

6 ☐☐☐☐☐☐☐
SAVINGS   DATE

NAME _____

ADDRESS _____

X _____
PLEASE SIGN IN TELLER'S PRESENCE FOR CASH RECEIVED

UNIT 6 ■ BANKING

4. On January 20, deposit checks for $15, $60.12, and $15.90. Receive $20 in cash.

Use the filled-in deposit slip to answer Problems 1-6.

1. How much cash was deposited (paper money and coins combined)? _____

2. How many checks were deposited? _____

3. What is the total amount in checks that was deposited? _____

4. How much cash was received by Fujio Tokuda? _____

5. What is the total amount that will be credited to this checking account? _____

6. List any combination of coins that would equal $2. _____

184   UNIT 6 ■ BANKING

7. Fill out the front and back of the deposit slip using the following information.

   A. Deposit $7 in currency.
   B. Deposit 75¢ in coin.
   C. Deposit checks for $3.52, $130, $23.50, $8, $13.47, and $10.
   D. Receive $100 in cash and sign your name.
   E. Write the net deposit.

PLEASE LIST EACH CHECK SEPARATELY AND SPECIFY BANK ON WHICH EACH IS DRAWN

ADDITIONAL CHECKS

	DOLLARS	CENTS
4		
5		
6		
7		
8		
9		
10		
11		
12		
13		
14		
15		
16		
17		
18		
19		
TOTAL		

ENTER THIS TOTAL ON FRONT

ALL CHECKS CREDITED SUBJECT TO FINAL PAYMENT

DIABLO BANK
2730 SALVIO STREET
CONCORD, CALIFORNIA 94519

CHECKING ACCOUNT DEPOSIT TICKET

DATE _____ 19 ____

SIGN HERE FOR LESS CASH IN TELLER'S PRESENCE

CASH — CURRENCY / COIN
LIST CHECKS SINGLY

TOTAL FROM OTHER SIDE
TOTAL
LESS CASH RECEIVED
NET DEPOSIT

USE OTHER SIDE FOR ADDITIONAL LISTING

BE SURE EACH ITEM IS PROPERLY ENDORSED

⑆021012349⑆057↿032118074⑈

UNIT 6 ■ BANKING

185

Writing Checks

When you write a check, you are telling your bank to give money in your account to the person or company to whom you wrote the check.

The following steps explain how to write a check. Study the example as you read the steps.

1. Always write in ink. Do not use an erasable pen or pencil.
2. Do not erase or cross out. Tear up any check on which you make a mistake.
3. Write the date you are actually writing the check. Checks dated on Sundays are just as good as those dated on other days.
4. Fill in the name of the person or company to whom you are writing the check.
5. Write the amount of the check in two ways: once using all numbers and once using words and numbers. If the amounts are different, the written amount is paid by the bank.
6. When you write the money amount in numbers, start as close to the dollar sign as possible so the amount can't be changed. When you write the money amount in words, start writing as far to the left as possible for the same reason. Capitalize the first word. Draw a line to fill any remaining space so that the amount can't be changed.
7. Sign the check. Your signature on the check must match your signature on the signature card on file at the bank.

Example

Answer the following questions by looking at the example.

1. Who signed the check? _____
2. To whom is the check written? _____
3. For what amount is this check written? _____
4. What number check is this? _____

Compare the two checks below. Then answer the questions that follow.

```
┌─────────────────────────────────────────────────────────────┐
│              LEE SHEAFFER                            615    │
│              PH. 555-6732                                   │
│           509 WASHINGTON STREET     Dec. 15  19 88  01-1234 │
│              MELROSE, IL 60639                       0210   │
│   PAY TO THE   Pat Hartley                    $  1 57/100   │
│   ORDER OF                                                  │
│   One and 57/100                             _____ DOLLARS │
│                                 FOR CLASSROOM USE ONLY      │
│   FIRST MIDWEST BANK                                        │
│       520 STATE STREET                                      │
│    CHICAGO, ILLINOIS 60605          Lee Sheaffer            │
│   MEMO_____    _____ │
│   ⑆0210 1234 9⑆ 615⑈ 0321 180 748                           │
└─────────────────────────────────────────────────────────────┘
```

```
┌─────────────────────────────────────────────────────────────┐
│              LEE SHEAFFER                            615    │
│              PH. 555-6732                                   │
│           509 WASHINGTON STREET     Dec. 15  19 88  01-1234 │
│              MELROSE, IL 60639                       0210   │
│   PAY TO THE   Pat Hartley                    $  21 57/100  │
│   ORDER OF                                                  │
│   Twenty-One and 57/100                      _____ DOLLARS │
│                                 FOR CLASSROOM USE ONLY      │
│   FIRST MIDWEST BANK                                        │
│       520 STATE STREET                                      │
│    CHICAGO, ILLINOIS 60605          Lee Sheaffer            │
│   MEMO_____    _____ │
│   ⑆0210 1234 9⑆ 615⑈ 0321 180 748                           │
└─────────────────────────────────────────────────────────────┘
```

1. What changes did someone make on check #615? _____
   _____

2. Rewrite check #615 so that the dollar amount can't be changed.

```
┌─────────────────────────────────────────────────────────────┐
│              LEE SHEAFFER                            615    │
│              PH. 555-6732                                   │
│           509 WASHINGTON STREET     _____ 19__ 01-1234 │
│              MELROSE, IL 60639                       0210   │
│   PAY TO THE   _____   $_____  │
│   ORDER OF                                                  │
│   _____ DOLLARS      │
│                                 FOR CLASSROOM USE ONLY      │
│   FIRST MIDWEST BANK                                        │
│       520 STATE STREET                                      │
│    CHICAGO, ILLINOIS 60605      _____ │
│   MEMO_____    _____ │
│   ⑆0210 1234 9⑆ 615⑈ 0321 180 748                           │
└─────────────────────────────────────────────────────────────┘
```

UNIT 6 ■ BANKING

The table below shows the correct spelling for some number amounts. Write checks using the given information.

Spelling List for Check Amounts			
seven	thirteen	eighteen	fifty
eight	fourteen	nineteen	sixty
nine	fifteen	twenty	seventy
eleven	sixteen	thirty	eighty
twelve	seventeen	forty	ninety

1. Write check #1 to Riviera Heights Apts. for rent, $450.

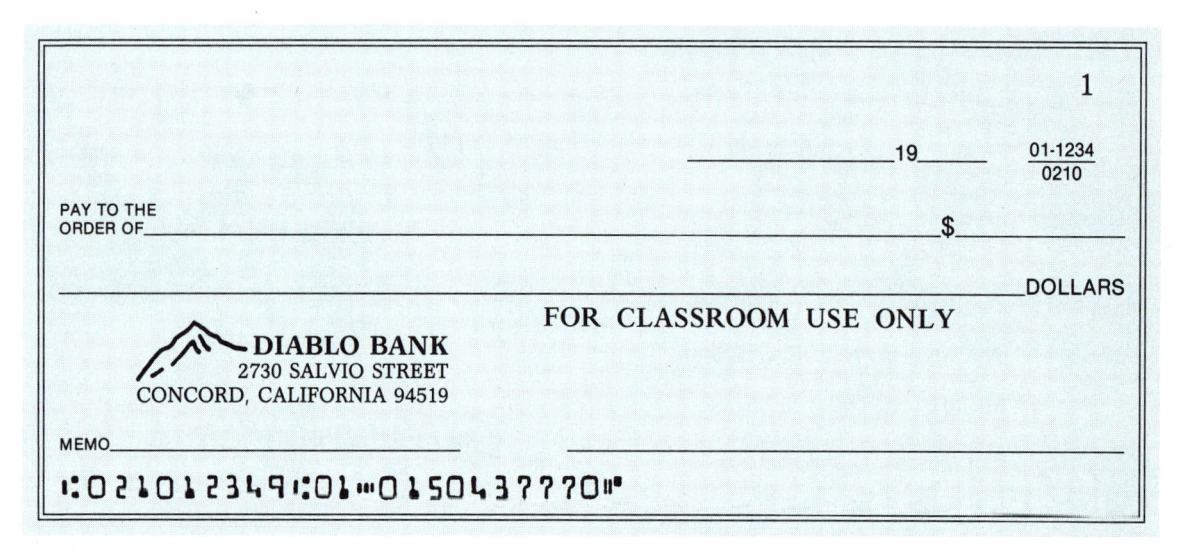

2. Write check #2 to Super Mart for groceries, $56.87.

188   UNIT 6 ■ BANKING

3. Write check #3 to Contra Costa Delivery for books, $15.95.

4. Write check #4 to Shoreway Drugs for prescriptions, $18.63.

5. Write check #5 to Midwest Record Store for albums, $17.50.

UNIT 6 ■ BANKING 189

6. Write check #6 to West States Telephone Co. for a phone bill, $47.24.

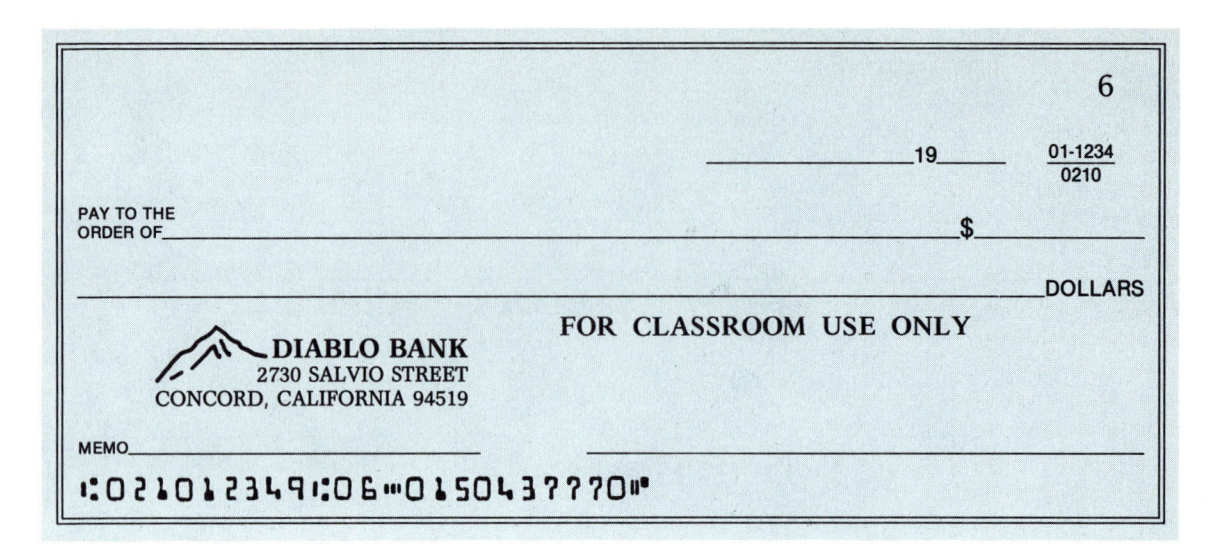

7. Write check #7 for $25. Make it payable to a friend (any name) for money you owe.

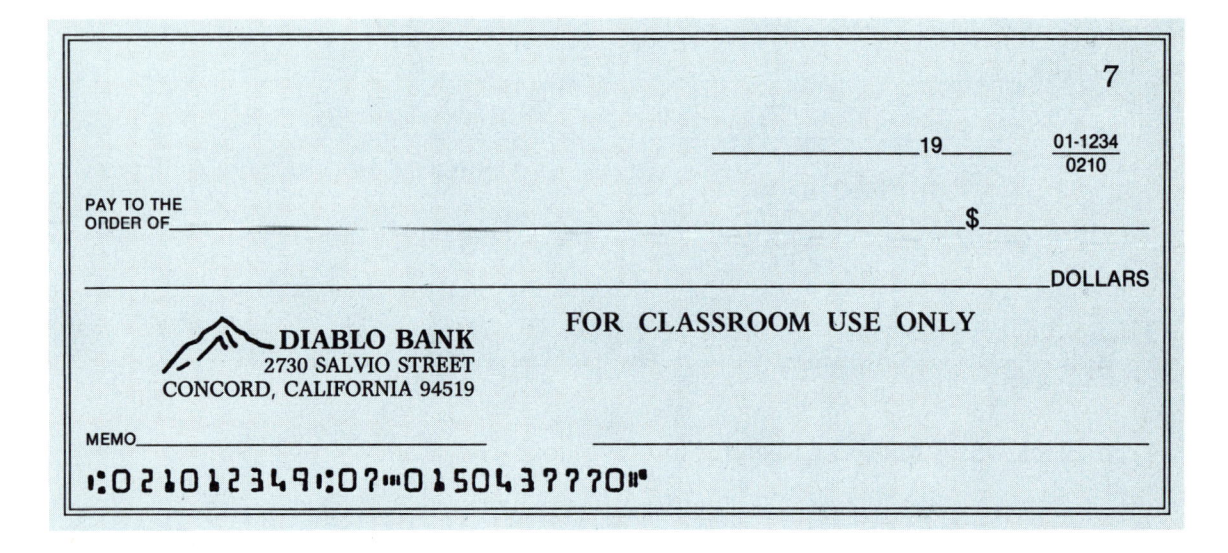

FILLING OUT A CHECK REGISTER

It is very important to keep an accurate record of the checks you have written and the deposits you have made. A check register is usually separate from your checks but fits in the same folder. (Some check registers are attached to the side of the checks.) If you keep accurate and up-to-date records, you will always know how much money is in your checking account.

A check register has places to record both checks and deposits. The following steps describe how to record checks in a check register. Study the example as you read the steps.

1. Write the number of the check in the column labeled *No.*
2. Enter the date the check was written in the column labeled *Date*.
3. Write to whom the check was made out on the line labeled *To*.
4. Write the item for which the check was written on the line labeled *For*.
5. Write the amount of the check in the column labeled *Subtractions*.
6. Also write the amount of the check in the column labeled *Balance Forward*. *Subtract* the amount of the check from the current balance. Then check your subtraction.

The following steps describe how to record deposits. Study the example as you read the steps.

1. Enter the date of the deposit in the *Date* column.
2. Write the word *Deposit* on the *To* line.
3. Write the amount of the deposit in the *Amount of Deposit or Interest* column.
4. Also write the amount of the deposit in the *Balance Forward* column. *Add* the amount of the deposit to the current balance. Check your addition.

Example

NO.	DATE	DESCRIPTION OF TRANSACTION	AMOUNT OF PAYMENT OR WITHDRAWAL (−)	✓	OTHER	AMOUNT OF DEPOSIT OR INTEREST (+)	BALANCE FORWARD
							149 85
184	5/8	TO Main Street Records FOR Cassette	7 27				− 7 27 142 58
185	5/10	TO F.O. Florists FOR Mother's Day Gift	15 25				−15 25 127 33
	5/15	TO Deposit FOR				55 00	+55 00 182 33
186	5/25	TO Cash FOR Weekend expenses	25 00				− 25 00 157 33
		TO FOR					
		TO FOR					

Answer the following questions by looking at the example.

1. How much money was in this account before check #184 was written? _____

2. On what day was a deposit made? _____

3. How much money was withdrawn for weekend expenses? _____

4. Which check was used to pay for a Mother's Day gift? _____

Use the information below to fill out the check register that follows.

May 3	You have a balance of $474.26. Write check #1 for $18.88 to Carson Dept. Store for a pair of shoes.
May 5	Write check #2 to Jim's Auto Body for a paint job, $275.
May 11	Write check #3 to King Hardware for lumber, $20.95.
May 15	Deposit $640.23.
May 16	Write check #4 for cash, $70, for a trip to the lake.
May 18	Write check #5 to Bruno's Car Repair, $86.50, for a car tune-up.
May 21	Write check #6 for $20.89 to Jeans for Everyone for a pair of jeans.
May 22	Write check #7 to Crown Station, $11.52, for gas.
May 23	Write check #8 for $20 to Bridget's Gifts for a wedding present.
May 23	Deposit $100.
May 25	Write check #9 to Peoria Veterinary Hospital for surgery on your dog's broken leg, $365.
May 27	Write check #10 to a friend (any friend) for a graduation gift, $20.

BANK STATEMENTS

Once a month your bank will send you a report of your account. This bank statement will list the checks paid, deposits made, and your new balance. It will also show any bank fees or charges that were subtracted from your account. These fees could include service charges, check printing charges, or fees for other bank services. You must *subtract* these fees or charges from your check register.

The parts of a bank statement are as follows:

1. *Account Number:* the same number as the account number on your checks
2. *Statement Date:* the period of time the statement includes
3. *Previous or Beginning Balance:* the amount in your account at the start of this statement
4. *Checks and Other Debits:* a listing of all checks or other charges that were debited (subtracted) from your account
5. *Deposits:* a listing of all deposits or additions to your account
6. *Ending Balance:* the amount left after all debits and deposits have been made
7. *Service Charge:* the amount of money the bank charged you for your checking account

An example of a checking account statement follows. There are 7 blank lines on the statement. Identify the parts of the statement by putting the numbers from the above list on the appropriate lines. Number 1 has been done for you.

DIABLO BANK

ELOISE GARCIA
14 RIDGE ROAD
PITTSBURGH, PA 15218

Account Number ___1___

Statement Period _____
1/10/88 THROUGH 2/09/88

SERVICES SUMMARY

Previous Statement Balance 1/10/88 .... $417.37
Deposits 275.32
Debits.. 510.40
Service Charge 5.00
New Balance... $177.29

DATE	CHECKS AND OTHER DEBITS	DEPOSITS AND OTHER CREDITS	BALANCE
			BEGINNING BALANCE _____
			1/10 $417.37
1/12	132 23.00 _____		394.37
1/13	133 5.00		389.37
1/26	136* 5.00		384.37
1/29	137 20.00		364.37
1/31		215.00 _____	579.37
2/02		60.32	639.69
2/05	138 272.25		367.44
2/06	139 29.15		338.29
	140 56.00		282.29
	141 100.00		182.29
2/09	5.00 SC _____		177.29

*GAP IN CHECK SEQUENCE

TOTAL DEBITS 510.40   TOTAL CREDITS 275.32   ENDING BALANCE 177.29 _____

UNIT 6 ■ BANKING

AUTOMATIC TELLER MACHINES (ATM)

Many banks allow customers access to their money through an *automatic teller machine (ATM)*. The ATM is located outside the bank and is connected to the bank's computer. By using an ATM, you can do your banking when the bank is closed, such as evenings or weekends.

Many banks participate in *Master-Teller*, a nationwide ATM system. This system allows a customer to withdraw cash at any Master-Teller system throughout the United States. There is a charge for using this system.

To use an ATM you need an identification card and a secret code number. These are issued by your bank. The secret code number is known only to you.

Although automatic teller machines look different, their services and operations are similar. The following services are available at an ATM:

1. Withdraw cash.
2. Make deposits.
3. Transfer funds between accounts.
4. Make payments to the bank for loans or credit card accounts.
5. Check account balances.

The following steps are used to operate an automatic teller machine.

1. Insert your plastic identification card.
2. Read the instructions on the viewer screen.
3. Enter your secret code number.
4. Select the transaction.
5. Enter the amount.
6. Read the amount on the viewer screen, and press OK or Cancel.
7. Receive cash or make a deposit.
8. Remove your card and your receipt.
9. Record your transactions in your check register.

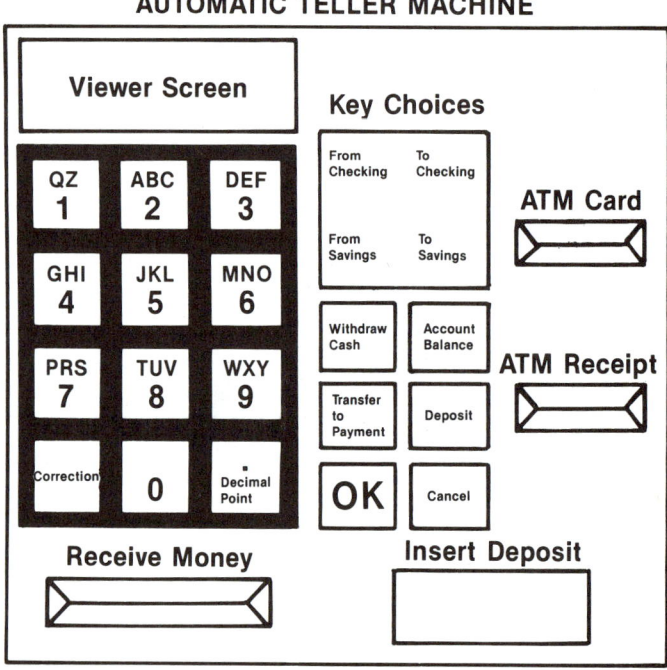

Answer the following questions.

1. If you lost your ATM card and someone else found it, could they get money from your account? Why or why not?

   _____

   _____

2. Why is it important to keep the receipt from your ATM transaction?

   _____

3. Study the ATM receipt at the right.

   A. What amount was withdrawn?

   _____

   B. What is the new balance?

   _____

   DIABLO BANK

ATM No.	Date	Time	Type	Trans No.
175E	07-16-87	10:02AM	01	09605

Customer No.	Location
392011607	CONCORD

Account Accessed	Amount	New Balance
01755700XX	$ 40.00	$ 864.13

   THANK YOU FOR USING EXPRESS

   TRANSACTION RECORD
   Please save for use in balancing your account.

4. Recording ATM transactions in your check register is essential in keeping an accurate balance of your account. Fill out the check register that follows using the information below.

   - Record a beginning balance of $87.
   - Record check #854 on June 1 to Cabletown for cable television fee, $13.14.
   - Record check #855 on June 5 to Gulf Telephone for May bill, $33.24.
   - Deposit a $1,300 check on June 10 at ATM.
   - Withdraw $200 from ATM on June 11.
   - Record check #856 on June 15 to Arnold Brown for rent, $450.

NO.	DATE	DESCRIPTION OF TRANSACTION	AMOUNT OF PAYMENT OR WITHDRAWAL (−)	✔	OTHER	AMOUNT OF DEPOSIT OR INTEREST (+)	BALANCE FORWARD
		TO					
		FOR					
		TO					
		FOR					
		TO					
		FOR					
		TO					
		FOR					
		TO					
		FOR					
		TO					
		FOR					

   A. What was the minimum balance during this time period? _____

   B. What might happen if you forgot to record an ATM transaction in your check register?

   _____

   _____

   C. Why might the ATM balance be larger than the balance indicated in your check register?

   _____

UNIT 6 ■ BANKING

CHECKING YOUR BALANCE

A bank statement is a record of all banking activity for a month. It is extremely important to carefully compare your check register with your bank statement. The statement and your check register should balance.

The following steps are used with a bank reconciliation worksheet to balance your check register with your bank statement.

1. In your check register, check off all cashed checks and all deposits shown on the bank statement.
2. Draw a line in your check register after the *date* of the last deposit or check recorded on the bank statement.
3. Subtract service charges from your check register.
4. Use the bank reconciliation worksheet to list outstanding checks (those not yet paid by the bank).
5. Balance your account by completing the bank reconciliation worksheet.

Following are a check register and bank statement for the month of September and a bank reconciliation worksheet. Use the bank statement and check register to complete the worksheet.

NO.	DATE	DESCRIPTION OF TRANSACTION	AMOUNT OF PAYMENT OR WITHDRAWAL (−)	✓	OTHER	AMOUNT OF DEPOSIT OR INTEREST (+)	BALANCE FORWARD 660 47
153	9/4	TO Luchas Market FOR Food	52 52				52 52 / 607 95
154	9/5	TO Valley Vet FOR	212 75				212 75 / 395 20
155	9/5	TO Luchas Market FOR Food	20 25				20 25 / 374 95
156	9/8	TO Aqua Compana FOR Water bill	47 22				47 22 / 327 73
157	9/10	TO Long Distance Service FOR Phone bill	30 14				30 14 / 297 59
158	9/11	TO Diablo Bank FOR charge card	226 00				226 00 / 71 59
	9/11	TO ATM FOR	40 00				40 00 / 31 59
	9/21	TO Deposit FOR				956 29	956 29 / 987 88
159	9/22	TO Local News FOR Paper	25 00				25 00 / 962 88
160	9/23	TO Cablevision FOR	29 44				29 44 / 933 44
161	9/23	TO VOID FOR					
162	9/28	TO Dress Rite FOR Pants	31 82				31 82 / 901 62
163	10/5	TO Mirrors Shoes FOR	100 00				100 00 / 801 62
		TO FOR					

196 UNIT 6 ■ BANKING

DIABLO BANK

MARCIE MILLER
205 RIVER LANE
HAMLET, MT 85211

Account Number 52311

Statement Period

9/03/89 THROUGH 10/03/89

SERVICES SUMMARY

Previous Statement Balance 9/03/89	$660.47
Deposits	956.29
Debits	690.14
Service Charge	5.00
New Balance	$921.62

DATE	CHECKS AND OTHER DEBITS	DEPOSITS AND OTHER CREDITS	BALANCE
			BEGINNING BALANCE
			9/03 $660.47
9/04	153 52.52		607.95
9/05	154 212.75		395.20
	155 20.25		374.95
9/08	156 47.22		327.73
9/10	157 30.14		297.59
9/11	158 226.00		71.59
	40.00 ATM		31.59
9/21		956.29	987.88
9/23	160* 29.44		958.44
9/28	162* 31.82		926.62
10/01	5.00 SC		921.62
	*GAP IN CHECK SEQUENCE		

TOTAL DEBITS 690.14   TOTAL CREDITS 956.29   ENDING BALANCE 921.62

Bank Reconciliation Worksheet

Outstanding Checks	
Number	Amount
Total	

ENTER new balance shown on statement $ _____

ADD deposits in check register not shown on statement   + $ _____

SUBTOTAL   = $ _____

SUBTRACT the outstanding items   − $ _____

BALANCE   = $ _____

Check Register Balance _____

Reconciliation Balance _____

Balance should agree with check register after service charge is subtracted from register.

UNIT 6 ■ BANKING

Use the bank statement below to answer the following questions.

DIABLO BANK

RONALD POWELL
1322 MONTAZUMA ST.
YUMA, AZ 85364

Account Number   46703389

Statement Period

5/08/89 THROUGH 6/10/89

SERVICES SUMMARY

Previous Statement Balance 5/08/89	$ 98.38
Deposits	1,800.00
Debits	1,854.40
Service Charge	5.00
New Balance	$ 38.98

DATE	CHECKS AND OTHER DEBITS	DEPOSITS AND OTHER CREDITS	BALANCE
			BEGINNING BALANCE
			5/08 $ 98.38
5/12	3074 50.00		48.38
5/13		1,200.00	1,248.38
5/14	3076* 40.00		1,208.38
5/17	3077 800.00		408.38
5/20	3078 35.00		373.38
5/21	3079 131.57		241.81
5/22		600.00	841.81
5/26	3082* 586.42		255.39
6/03	3083 126.32		129.07
6/08	3084 85.09		43.48
6/10	5.00 SC		38.98
	*GAP IN CHECK SEQUENCE		

TOTAL DEBITS 1,854.40   TOTAL CREDITS 1,800.00   ENDING BALANCE 38.98

1. What does the * by check numbers 3076 and 3082 mean?

   _____

2. Ron's check register shows that check #3075 for $27.30, check #3080 for $38.52, and check #3081 for $40 did not appear in his statement. (The bank has not yet subtracted them from his account.) What is the total amount of Ron's unpaid checks?   _____

   $27.30 + $38.52 + $40 = $105.82

3. On June 10, what amount of money was available for Ron to write checks?   _____

4. Give 3 examples of why a person might have a check "bounce" (not be paid by the bank) because there was not enough money in the account.

   A. _____

   B. _____

   C. _____

198   UNIT 6 ■ BANKING

7 GEOMETRIC SHAPES AND CALCULATIONS

SKILLS YOU WILL NEED

　　Basic math skills

SKILLS YOU WILL LEARN

　　Identifying geometric shapes
　　Finding perimeters
　　Finding circumferences of circles
　　Finding areas of rectangles, squares, and circles
　　Finding volumes of rectangular objects

SITUATIONS IN WHICH YOU'LL USE THESE SKILLS

　　In cooking
　　In sewing
　　In home decorating
　　In building projects

　　Geometric shapes are all around you. Think of your room at home. The floor, walls, ceiling, door, and windows are probably rectangles. You may have a round or a square table.

　　The ability to identify and measure geometric shapes is a skill you will often use in everyday life. This unit discusses basic geometric shapes and shows you how to find perimeter, circumference, area, and volume.

UNIT 7 ■ GEOMETRIC SHAPES AND CALCULATIONS

GEOMETRIC FIGURES

Different geometric figures are measured different ways. You need to be able to identify the geometric figure before you can measure it.

Two-dimensional figures are flat figures. The most common two-dimensional figures are the following.

Parallelogram   A parallelogram is a four-sided figure that has two pairs of parallel sides.

> **HINT:** Parallel lines are the same distance apart at every point. Parallel lines will never meet in a point.

Rectangle   A rectangle is a four-sided figure with parallel, opposite sides. The corners of a rectangle are right angles. A rectangle is a special kind of parallelogram.

> **HINT:** A right angle is like the corner of a sheet of paper. The ⌐ in the corner of the rectangles is used to show a right angle.

Square   A square is a four-sided figure with four equal sides. Each corner of a square is a right angle. A square is a special kind of rectangle.

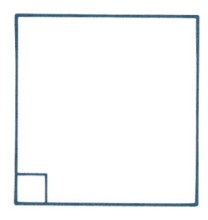

Triangle   A triangle is a three-sided figure.

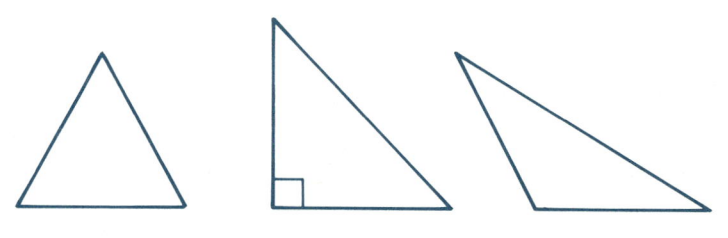

Circle A circle is a closed curved line, everywhere an equal distance from a given point called the *center*.

Oval An oval is an elongated, curved figure. This shape is also called an *ellipse*.

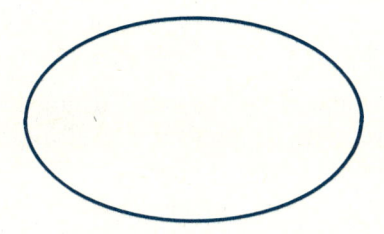

Three-dimensional figures have depth. The most common three-dimensional figures are the following.

Sphere A sphere has a rounded surface. Every point on the surface is the same distance from the center. A baseball is a sphere.

Cube A cube is a three-dimensional figure having equal length, width, and height. Sugar is often shaped as cubes.

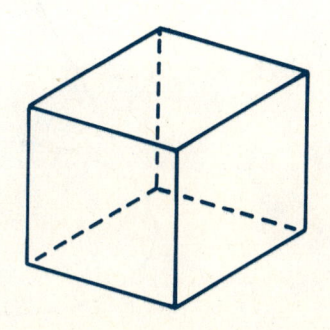

UNIT 7 ■ GEOMETRIC SHAPES AND CALCULATIONS

Cylinder A cylinder is a three-dimensional figure with straight sides. The top and bottom of a cylinder are two equal and parallel circles. A can of soup is an example of a cylinder.

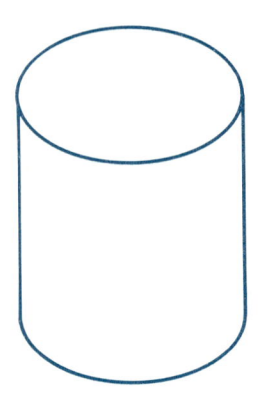

Parallelograms, rectangles, squares, and triangles are polygons. A *polygon* is a flat figure with sides that are straight lines. There are many different kinds of polygons. For example, a stop sign is a polygon with eight sides.

Pyramid A pyramid has a polygon for its base and triangular faces that meet at a point. The most common pyramids are those with either a square base or a triangular base.

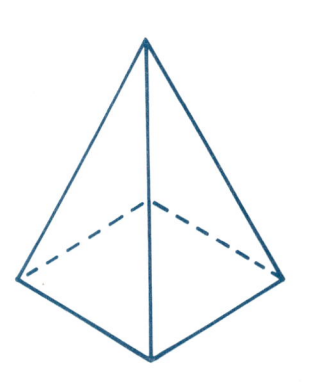

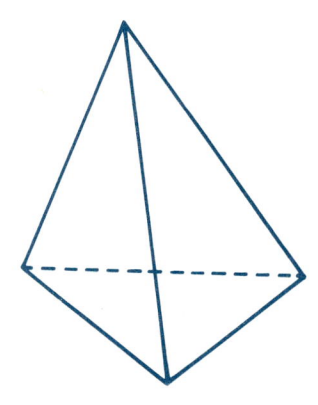

Cone A cone has a circle for its base and curved sides that taper evenly to a point. The cone of an ice cream cone is an example.

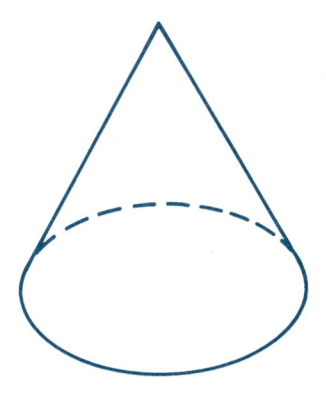

Answer the following questions about geometric shapes.

1. Which of the following is most like a cylinder? _____

   A. An egg
   B. The sun
   C. A can of beans
   D. A straight pin

2. If a recipe calls for a square pan, which one would you use? _____

   A. (circle)
   B. (triangle)
   C. (rectangle)
   D. (square)

3. Which of the following is most like a sphere? _____

   A. A dish
   B. An orange
   C. A spoon
   D. A football

4. If a pattern calls for a circular piece, which one would you use? _____

   A. (square)
   B. (ellipse)
   C. (circle)
   D. (rectangle)

5. Name the geometric shape for each of the following objects.

   A. Soccer ball _____
   B. Turkey platter _____
   C. Tepee _____
   D. Typing paper _____
   E. Can of peaches _____
   F. Cardboard box _____
   G. Book cover _____
   H. Record album cover _____
   I. Basketball hoop _____
   J. Orange _____

UNIT 7 ■ GEOMETRIC SHAPES AND CALCULATIONS

PERIMETER

Perimeter is the distance around the outside of a two-dimensional figure. To find the perimeter of an object, add the lengths of all the sides.

Example 1   Find the perimeter of the following rectangle.

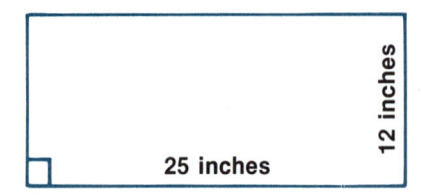

```
        25 inches
      12 inches
    + 12 inches
      74 inches
```

25 inches
25 inches
12 inches
+ 12 inches
74 inches

Answer: The perimeter is 74 inches.

In Example 1, the sides opposite each other are of equal length. There are two 25-inch sides and two 12-inch sides. Another way to solve this problem would be as follows.

2 × 25 inches = 50 inches   50 inches
2 × 12 inches = 24 inches   + 24 inches
                                 74 inches

Example 2   Find the perimeter of the following figure.

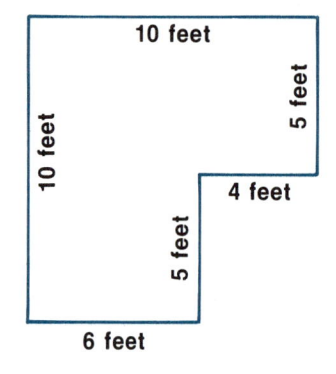

Add the lengths of the 6 sides.

10 feet
5 feet
4 feet
5 feet
6 feet
+ 10 feet
40 feet

Answer: The perimeter is 40 feet.

Solve the following problems.

1. Find the perimeter of the rectangle.   _____

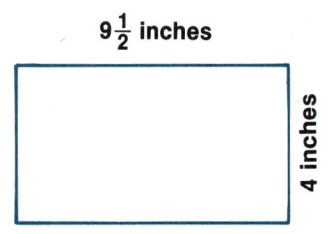

   $9\frac{1}{2}$ inches
   4 inches

204   UNIT 7 ■ GEOMETRIC SHAPES AND CALCULATIONS

2. Find the perimeter of the triangle.

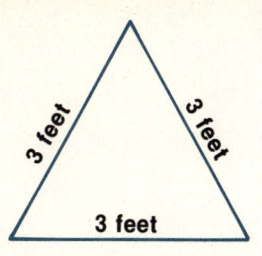

   (Triangle with sides 3 feet, 3 feet, 3 feet)

3. How much tape would be needed to go around the outside of a window that is $18\frac{1}{2}$ inches long and $12\frac{1}{4}$ inches wide?

If you know the perimeter and the measurement of one side of a square or rectangle, you can find the measurements of the other sides.

Example 3   The perimeter of the rectangle is 36 inches. The width (the short side) is 7 inches. What is the measurement of the long side (the length)?

Step 1   Add the measurements of the two short sides (the widths).

7 inches + 7 inches = 14 inches

Step 2   Subtract this sum from the perimeter.

36 inches − 14 inches = 22 inches

Step 3   Divide the answer in Step 2 by 2.

22 inches ÷ 2 = 11 inches

Answer: The length is 11 inches.

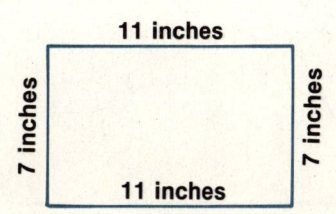

UNIT 7 ■ GEOMETRIC SHAPES AND CALCULATIONS

Answer the following questions.

1. Find the length of a patio that has a width of $11\frac{1}{2}$ feet, and a perimeter of 59 feet. _____

2. Find the width of a bedspread that has a length of 108 inches and a perimeter of 396 inches. _____

3. Stuart has purchased 36 feet of lumber to build a rectangular sandbox. If he wants the width to be 6 feet, what will the length be? _____

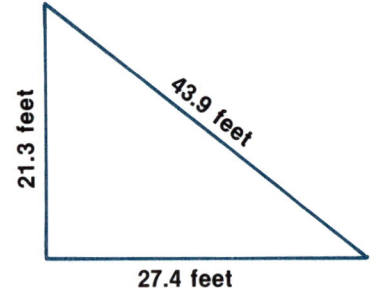

4. Find the perimeters of the following two figures.

    A. _____

    (triangle with sides 21.3 feet, 43.9 feet, and 27.4 feet)

    B. _____

    (parallelogram with sides $37\frac{3}{4}$ inches and $27\frac{2}{3}$ inches)

206   UNIT 7 ■ GEOMETRIC SHAPES AND CALCULATIONS

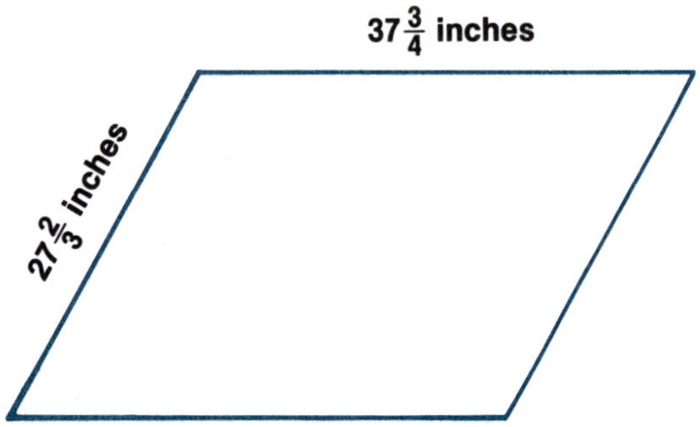

5. Louise has made an embroidery wall hanging that is 3 feet by 5 feet. How many feet of wood will she need to make a frame? _____

6. How many feet of trim will be necessary to edge a tablecloth that measures 87 inches by 54 inches? _____

7. Glen's rectangular garden needed 104 feet of fencing. If the garden is 31 feet long, how wide is it? _____

8. Aaron is making a set of 4 rectangular place mats. Each place mat measures 18 inches by 24 inches. How many feet of trim will he need to edge all the place mats? _____

9. Corinne is planning to fence in her vegetable garden that is 48 feet long and 16 feet wide. How many feet of fencing will she need to buy? _____

UNIT 7 ■ GEOMETRIC SHAPES AND CALCULATIONS

10. Al is putting weather stripping in his house. Each window is $2\frac{1}{2}$ feet wide and $4\frac{1}{4}$ feet high. There are 7 windows. How many feet of weather stripping will he need for all 7 windows? _____

Circumference

The perimeter of a circle is called its *circumference*. To find the circumference of a circle you have to know the radius or the diameter of the circle.

The *diameter* of a circle is the distance across the circle through the center.

The *radius* of a circle is the distance from the center to the outside of the circle.

> **HINT:** The diameter is 2 times the radius.

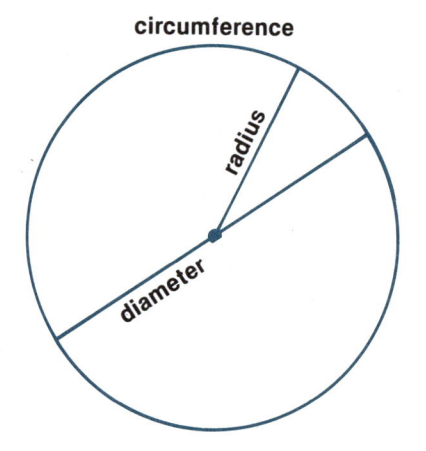

The ratio of circumference to diameter is the same for all circles. The letter pi (π), from the Greek alphabet, is used as the symbol for this ratio. The approximate value of pi is 3.14 or $\frac{22}{7}$.

To find the circumference (C) of a circle, multiply pi (π) times the diameter (d). Another way to find the circumference is to multiply 2 times π times the radius (r).

> **Formula for Circumference**
>
> $C = \pi d$ or $C = 2\pi r$

> **HINT:** In all formulas, letters and symbols in succession with no mathematical symbol between them are to be multiplied.

Example 1 If the *diameter* of a circle is 8 inches, what is the circumference?

Step 1 Decide which formula to use. The diameter is given, so use C = πd.

Step 2 Substitute the known values in the formula and solve.

C = πd
C = 3.14 × 8 inches
C = 25.12 inches

Answer: The circumference is 25.12 inches.

Example 2 If the *radius* of a circle is 9 inches, what is the circumference?

C = 2πr
C = 2 × 3.14 × 9 inches
C = 56.52 inches

Answer: The circumference is 56.52 inches.

Answer the following questions. Use 3.14 for the value of π.

1. Identify the parts of the circle.

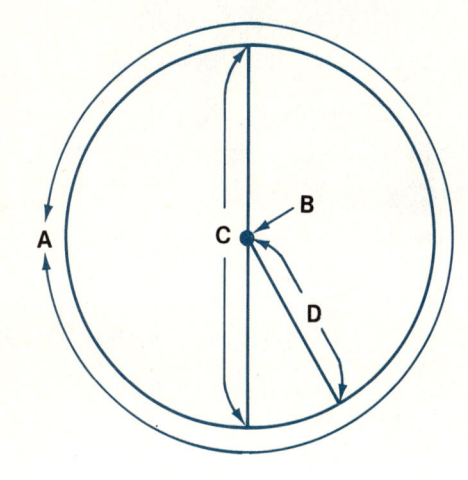

A. _____

B. _____

C. _____

D. _____

2. Find the circumferences of these circles.

A. 15 inches

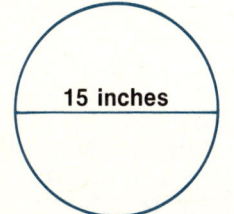

B. 29 inches

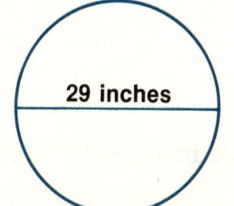

C. 7 inches

UNIT 7 ■ GEOMETRIC SHAPES AND CALCULATIONS

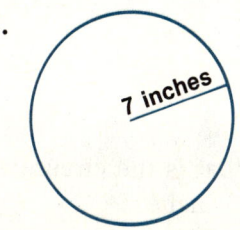

3. Myrna is making a square-dance skirt and wants to put trim around the bottom. She knows that the radius of the skirt is 23 inches. How many inches of trim will she need? _____

4. Sharon wants to buy snow chains for her tires. Her tires are 21 inches in diameter. What length chains must she buy?

_____

5. Perry made a circular stained glass window. The diameter of the window is 25 inches. What is the circumference of the wooden frame he will need to make? _____

6. Eileen is making 6 coasters. Each coaster is 3 inches in diameter. How much trim tape will she need to go around all the coasters? _____

7. What is the circumference of a water pipe that has a radius of 2.5 inches? _____

Note: Extra material may be required for mitring frame corners or for overlapping trim.

8. A preschool class is decorating empty coffee cans to hold crayons. What length tape is needed to go around a can having a 4-inch diameter? _____

9. Leo is making 2 circular flower beds that are each 4 feet in diameter. He wants to put an edging strip around each flower bed. How much edging strip should he buy for the flower beds? _____

AREA

Area is the measurement of space contained in a flat figure. Area is measured in square units such as square inches, square feet, or square meters.

Sometimes it is necessary to change square inches to square feet or to change square feet to square yards. Equivalent square units are as follows.

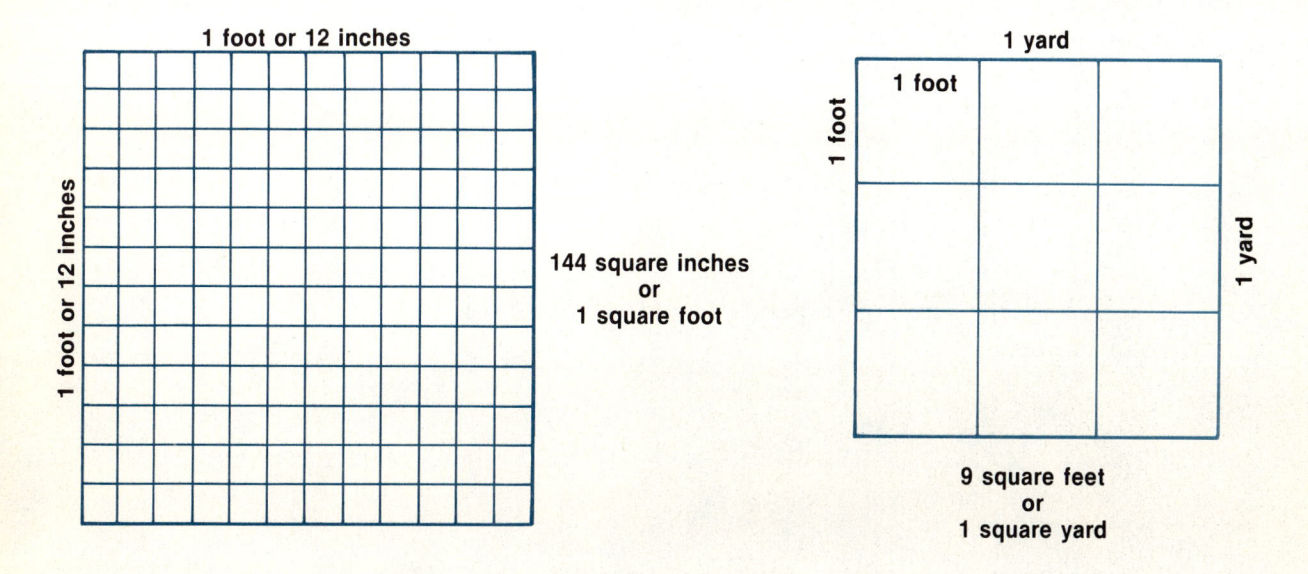

1 foot or 12 inches

1 foot or 12 inches

144 square inches
or
1 square foot

1 yard

1 foot

1 yard

9 square feet
or
1 square yard

UNIT 7 ■ GEOMETRIC SHAPES AND CALCULATIONS

Area of Rectangles and Squares

To find the area (A) of a rectangle or square, multiply the length (l) times the width (w).

> **Formula for Area of Rectangle or Square**
>
> A = lw

Example 1

Find the area of the following rectangle.

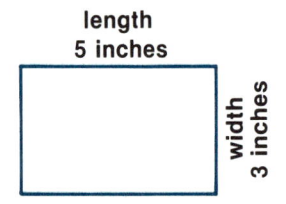

length
5 inches

width
3 inches

A = lw
A = length × width
A = 5 inches × 3 inches
A = 15 square inches

Example 2

Find the area of the following square.

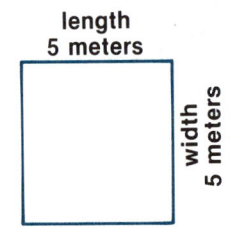

length
5 meters

width
5 meters

A = lw
A = length × width
A = 5 meters × 5 meters
A = 25 square meters

Answer the following questions.

1. What is the area of a room 13 feet wide and 18 feet long? _____

2. How much carpeting should you buy for a room that is 12 feet by 15 feet? Answer in square feet. _____

3. How many square yards of carpeting would you need for the room in Problem 2? _____

4. Becky wants to make a patchwork quilt 4 feet by 6 feet. Each patchwork piece is 1 square foot. How many pieces will be needed? _____

If you know the area and the measurement of one side of a square or rectangle, you can find the measurement of the other side by dividing the known side into the given area.

Example 3   What is the width of a rectangle that has a length of 20 feet and an area of 240 square feet?

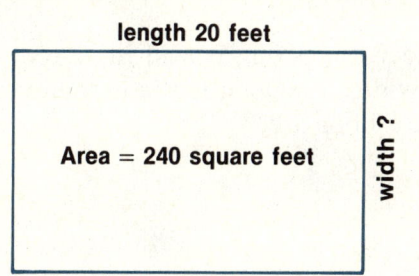

Divide the area by the known measurement.

```
      12
20)240
      20
      ‾‾
      40
      40
      ‾‾
       0
```

Answer: The width is 12 feet.

Answer the following questions.

1. What is the width of a room that has a length of 23 feet and an area of 368 square feet? _____

2. One gallon of paint will cover 288 square feet. If the room to be painted is 12 feet high, what is the length that can be covered with 1 gallon of paint? _____

UNIT 7 ■ GEOMETRIC SHAPES AND CALCULATIONS

3. What is the length of a motorcycle trailer that has a floor area of 32 square feet and a width of 4 feet? _____

4. Find the length of a rectangle that has a width of 5 feet and an area of 65 square feet. _____

5. A room measures 192 square feet. Carpeting is sold in 12-foot widths. If the room is 12 feet wide, how long is the carpeting you will need to cover the entire floor area? _____

6. Simon and Art used 1 gallon of paint to paint the playground fence. One gallon of paint covers 300 square feet. If the fence is 12 feet high, what is the length of the fence? _____

7. Annie is making a patchwork quilt. The finished size is to be 96 inches by 108 inches. How many finished 4-inch squares will she need? _____

8. Miss Powell is buying carpet for her living room, which measures 17 feet by 12 feet, and her hall, which measures $7\frac{1}{2}$ feet by $4\frac{1}{2}$ feet. How many square feet of carpet will she need? _____

9. The area of a play yard is 18,750 square feet. The length is 250 feet. How many feet is the width? _____

10. The Moore family is planning to build a deck in front of their living room. If the deck measures $15\frac{1}{2}$ feet by 8 feet, what is the area of the deck? _____

11. The wall of a convention room measures 10 feet by 25 feet. There are 2 windows, each measuring 4 feet by 6 feet. How many square feet will be paneled? (Remember not to panel over the windows.) _____

UNIT 7 ■ GEOMETRIC SHAPES AND CALCULATIONS

12. Roger wants to put tiles on the kitchen floor. Each tile is 1 square foot. How many tiles will he need, if the floor measures 12 feet by 31 feet? _____

13. A roll of wallpaper covers 32 square feet. How many rolls will you need for 2 walls that measure 8 feet by 10 feet and 8 feet by 12 feet? (Don't worry about matching the pattern.) _____

Area of Circles

To find the area (A) of a circle, multiply pi (π) times the square of the radius (r^2).

Formula for Area of Circle

$$A = \pi r^2$$

HINT: To square the radius, multiply radius times radius.

(5 inches)^2 = 5 inches × 5 inches = 25 square inches

Example   What is the area of a circle that has a radius of 5 inches?

$A = \pi r^2$
$A = 3.14 \times 5 \text{ inches} \times 5 \text{ inches}$
$A = 3.14 \times 25 \text{ square inches}$
$A = 78.5 \text{ square inches}$

Answer: The area of the circle is 78.5 square inches.

Answer the following questions. Use 3.14 for the value of π.

1. Find the areas of the following circles.

   A. ⊙ 28 feet _____

   B. ⊙ 13 inches _____

2. A circular mirror has a radius of 14 inches. What is the area of the mirror? _____

3. A garden sprinkler waters a distance, or diameter, of 21 feet. How many square feet can it water at one time? _____

4. The diameter of a circular cake pan is 10 inches. What is its area? _____

UNIT 7 ■ GEOMETRIC SHAPES AND CALCULATIONS 217

VOLUME

Volume is the amount of space inside a three-dimensional object. Volume is measured in cubic units, such as cubic inches, cubic feet, or cubic meters.

To find the volume (V) of a rectangular object, multiply the length (l) times the width (w) times the height (h).

> **Formula for Volume of Rectangular Object**
>
> V = lwh

Example 1   What is the volume of the rectangular box below?

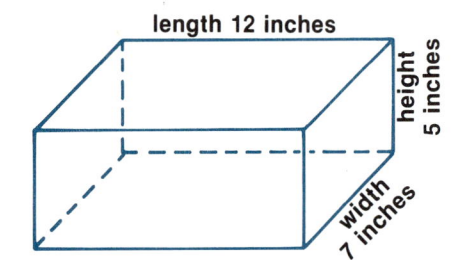

length 12 inches
height 5 inches
width 7 inches

V = lwh
V = 12 inches × 7 inches × 5 inches
V = 420 cubic inches

Answer: The volume of the rectangular box is 420 cubic inches.

Example 2   Find the volume of the following figure.

7 inches
7 inches
7 inches

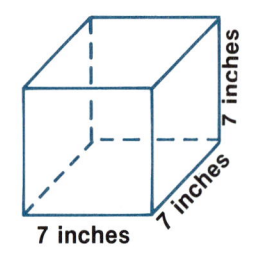

V = lwh
V = 7 inches × 7 inches × 7 inches
V = 343 cubic inches

Answer: The volume of the figure is 343 cubic inches.

Answer the following questions.

1. Find the volumes of the following figures.

   A. 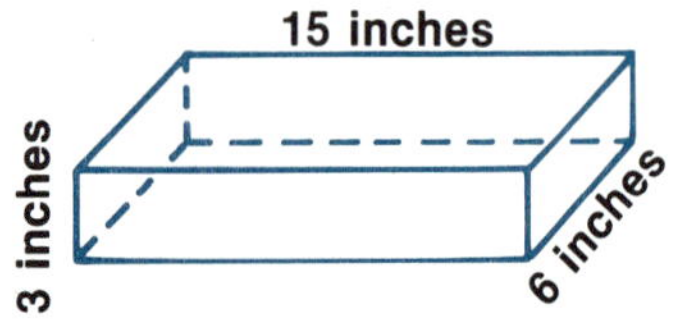 15 inches, 3 inches, 6 inches

   B. 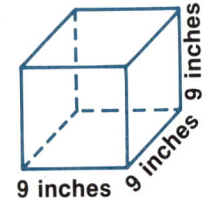 9 inches, 9 inches, 9 inches

   C. 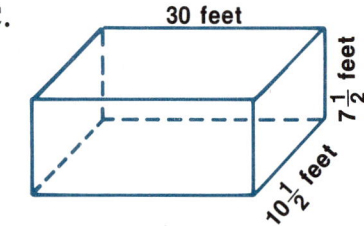 30 feet, $10\frac{1}{2}$ feet, $7\frac{1}{2}$ feet

---

218   UNIT 7 ■ GEOMETRIC SHAPES AND CALCULATIONS

2. If a storage area measures 54 inches high, 30 inches long, and 24 inches wide, how many cubic inches are in the storage area? _____

3. Ms. Matamoroz's pool is 15 meters long, 12 meters wide, and 3 meters deep. How many cubic meters of water does it hold? _____

4. Earl is digging a foundation that is 16 feet long, 12 feet wide, and 4 feet deep. How many cubic feet of soil will have to be removed? _____

5. Ingrid's aquarium measures 12 inches by 15 inches by 12 inches. How many cubic inches of water will fill the aquarium to the top? _____

6. Tracy wants to set her stereo speakers into a bookcase cabinet. How many cubic inches of space will each speaker take if one speaker measures 7 inches by 12 inches by 18 inches? _____

UNIT 7 ■ GEOMETRIC SHAPES AND CALCULATIONS

7. Molly and Matt want to build a fruit cellar in the backyard. They want the cellar to measure 10 meters by 12 meters by 8 meters. How many cubic meters of dirt do they need to remove to build the cellar? _____

8. Craig and his father plan to build a waterproof luggage carrier for the top of the family car. If the carrier measures $16\frac{1}{2}$ inches high, 41 inches wide, and $48\frac{1}{2}$ inches long, what is the volume of the luggage carrier? _____

8 TIME

SKILLS YOU WILL NEED

    Basic math skills
    Ability to read tables

SKILLS YOU WILL LEARN

    Figuring time intervals and conversions
    Recording hours worked on a time card
    Computing time and a half
    Figuring an hourly percent raise
    Using transportation schedules
    Using time zones

SITUATIONS IN WHICH YOU'LL USE THESE SKILLS

    Traveling
    Figuring weekly pay
    Planning your social calendar
    Being on time for appointments
    Planning meals

    Time is relative. It goes by very slowly when you're waiting for someone. And it slips by quickly when you're having fun. This unit will help you interpret and understand the everyday use of time.

FIGURING TIME INTERVALS

To figure out time intervals, you may need to convert units of time. The box at the right shows time conversions.

> 60 seconds = 1 minute
> 60 minutes = 1 hour
> 24 hours = 1 day
> 7 days = 1 week
> *365 days = 1 year
> 52 weeks = 1 year
> 12 months = 1 year
>
> *Leap years have 366 days.

Hours and Minutes

There are 24 hours in a day, but only 12 hours on a clock. The new day begins at midnight and ends 24 hours later at midnight. We use A.M. following a time to show the morning hours (midnight to noon). We use P.M. following a time to show the afternoon and evening hours (noon to midnight). There are 60 minutes in every hour.

Some clocks are numbered from 1 to 12. Between each of the 12 numbers are 5 minutes. The small hand of the clock points to the hour and the large hand points to the minute.

Another type of clock is the digital clock. These clocks are easy to read. The time is the numbers you see.

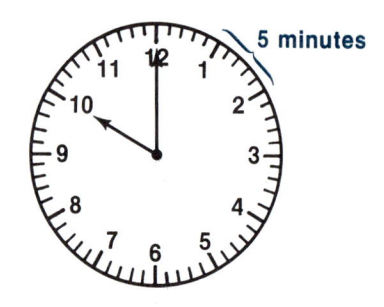

The time is 10:00.

To figure the number of whole hours between one time and another time, you can count the hours on your fingers or in your head.

Example 1 How many hours are there from 5:00 P.M. to 8:00 P.M.?

5:00 to 6:00 = 1 hour
6:00 to 7:00 = 1 hour
7:00 to 8:00 = 1 hour
            3 hours

5:00 P.M.   8:00 P.M.

Answer the following questions.

1. Mia watched the late-late show from 11 P.M. to 3 A.M. How many hours was she watching the show? _____

2. Seth and Pam, left their children with a babysitter from 5 P.M. until 1 A.M. For how many hours should they pay the babysitter? _____

222   UNIT 8 ■ TIME

To figure parts of an hour, count the number of minutes from starting time to ending time. It is easy to look at a clock and count the minutes in 5-minute groups.

Example 2 Greta got home at 2:05. She changed clothes and left at 2:26 for the concert. How long did it take Greta to get ready?

2:05   2:26

Answer: It took Greta 21 minutes to get ready.

Answer the following questions.

1. Ray worked on his math homework from 8:30 P.M. to 9:20 P.M. How long did he spend on his homework?

   _____

2. Gina waited for a friend from 10:15 A.M. to 10:37 A.M.. How long did she wait?

3. The twins swam in the river from 2:45 P.M. until 3:25 P.M. How long were they in the water?

4. The fire alarm went off at 1:33 A.M., and the fire truck arrived at 1:56 A.M. How long did it take for the firefighters to arrive at the fire?

5. A recipe calls for a cake to bake for 45 minutes. Norman put the cake in the oven at 9:10 A.M. When will the cake be done?

Minutes are sometimes written as a fraction of an hour. Usually only $\frac{1}{4}$ hour, $\frac{1}{2}$ hour, and $\frac{3}{4}$ hour are written as fractions.

15 minutes = $\frac{15}{60}$ = $\frac{1}{4}$ hour

30 minutes = $\frac{30}{60}$ = $\frac{1}{2}$ hour

45 minutes = $\frac{45}{60}$ = $\frac{3}{4}$ hour

HINT: Instead of saying there are 15 minutes between 12:30 P.M. and 12:45 P.M., we can say there is $\frac{1}{4}$ hour between the two times.

UNIT 8 ■ TIME

Answer the following questions. State your answers using fractions.

1. The muffins baked from 10:45 A.M. until 11 A.M. How long did they bake? _____

2. The laundry dried in the dryer from 7:16 P.M. until 8:01 P.M. How long did it take to dry? _____

3. Evelyn jogged from 6:20 A.M. until 6:50 A.M.. How long did she jog? _____

4. Greg was in the shower from 7:47 A.M. until 8:02 A.M. How long did he shower? _____

To figure a combination of minutes and hours, count both the hours and minutes.

Example 3   The football game started at 1:25 P.M. and ended at 4:15 P.M. How long did the game take?

From 1:25 P.M. to 3:25 P.M. is 2 hours.
From 3:25 P.M. to 4:15 P.M. is 50 minutes.

Answer: The football game took 2 hours 50 minutes.

Answer the following questions.

1. Patrick and Howard drove from San Francisco to Sacramento. They left San Francisco at 6:45 P.M. and arrived in Sacramento at 8:25 P.M. How long did the trip take? _____

2. Harold's dog was missing from 5:30 A.M. until 11:42 A.M. How long was the dog gone? _____

3. The rock band practiced from 5:45 P.M. until 1:23 A.M. How long were the neighbors entertained?

_____

4. It took Erica from 9:09 P.M. until 2:37 A.M. to finish sewing her dress for the dance. How long did she work? _____

Months and Days

S M T W T F S	S M T W T F S
JANUARY	**JULY**
1 2 3	1 2 3
4 5 6 7 8 9 10	4 5 6 7 8 9 10
11 12 13 14 15 16 17	11 12 13 14 15 16 17
18 19 20 21 22 23 24	18 19 20 21 22 23 24
25 26 27 28 29 30 31	25 26 27 28 29 30 31
FEBRUARY	**AUGUST**
1 2 3 4 5 6 7	1 2 3 4 5 6 7
8 9 10 11 12 13 14	8 9 10 11 12 13 14
15 16 17 18 19 20 21	15 16 17 18 19 20 21
22 23 24 25 26 27 28	22 23 24 25 26 27 28
29	29 30 31
MARCH	**SEPTEMBER**
1 2 3 4 5 6	1 2 3 4
7 8 9 10 11 12 13	5 6 7 8 9 10 11
14 15 16 17 18 19 20	12 13 14 15 16 17 18
21 22 23 24 25 26 27	19 20 21 22 23 24 25
28 29 30 31	26 27 28 29 30
APRIL	**OCTOBER**
1 2 3	1 2
4 5 6 7 8 9 10	3 4 5 6 7 8 9
11 12 13 14 15 16 17	10 11 12 13 14 15 16
18 19 20 21 22 23 24	17 18 19 20 21 22 23
25 26 27 28 29 30	24 25 26 27 28 29 30
	31
MAY	**NOVEMBER**
1	1 2 3 4 5 6
2 3 4 5 6 7 8	7 8 9 10 11 12 13
9 10 11 12 13 14 15	14 15 16 17 18 19 20
16 17 18 19 20 21 22	21 22 23 24 25 26 27
23 24 25 26 27 28 29	28 29 30
30 31	
JUNE	**DECEMBER**
1 2 3 4 5	1 2 3 4
6 7 8 9 10 11 12	5 6 7 8 9 10 11
13 14 15 16 17 18 19	12 13 14 15 16 17 18
20 21 22 23 24 25 26	19 20 21 22 23 24 25
27 28 29 30	26 27 28 29 30 31

There are 12 months in a year. The number of days in a month varies as shown on the calendar. It is important to know the months in order from January to December. Say the months as you count them on your fingers.

JAN, FEB, MAR, APR, MAY, JUN, JUL, AUG, SEP, OCT, NOV, DEC

Example 1   Candy will be playing basketball during the months of November through March. How many months will she be playing basketball?

November	1 month
December	1 month
January	1 month
February	1 month
March	1 month
	5 months

HINT: The word *through* tells you to count both November *and* March.

Answer: Candy will be playing basketball 5 months.

Answer the following questions.

1. Clothing students are required to complete 4 projects in a semester. The semester starts in September and ends in January. How many months does a semester include? _____

2. Peggy is planning to go on a cruise in 7 months. It is now August 15. In what month is the cruise? _____

UNIT 8 ■ TIME

225

Finding the number of days from one date to another can be very confusing. As the following example shows, there can be three different answers depending on the interpretation of the question.

Example 2   Cathy and Milt arrived at the lake at noon on February 3. They left at noon on February 6. How many days were they there?

They were at the lake for three 24-hour days.

They were at the lake four days: February 3, 4, 5, and 6.

They were at the lake all day on February 4 and 5.

Answer the following questions. Read each question carefully to decide whether the given dates are both included, whether only one is included, or whether the question asks for the number of days between the given dates.

1. Due to fire danger, Mountain Park was closed from September 13 through September 27. How many days was the park closed? _____

2. Chester will be leaving for New York on August 14. He will return home on August 26. How many days will he be away from home? _____

3. Vern and Lori are planning their wedding for December 24. If today is December 3, how many days do they have until the wedding day? (What are the different answers that could be given for this problem?) _____

4. Carrie starts work on Tuesday and works every day through Thursday of the next week. How many days does she work before she has a day off? _____

5. Eve started her summer sewing project on June 14 and finished it on July 3. How many days did she work on it if she worked every day? (June has 30 days.) _____

Use your knowlege of time intervals to answer the following questions.

1. Write the 12 months of the year in order.

   A. _____   G. _____

   B. _____   H. _____

   C. _____   I. _____

   D. _____   J. _____

   E. _____   K. _____

   F. _____   L. _____

2. How many days are in a regular year? _____

3. Use the bill below to answer the questions that follow.

Kindly Write Your Account Number on Face of Check or Money Order Made Payable to:	YORKS		AMOUNT ENCLOSED $	
JOYCE HARDY 289 3RD AVENUE WALNUT CREEK, CA 94596		YORKS Louisville, KY 40285-8440	MINIMUM PAYMENT DUE	29.00

   ACCOUNT NUMBER: 455 39 555

DATE	STORE	REFERENCE #	DEPT. #	MERCHANDISE AND TRANSACTION DESCRIPTION	PURCHASES	PAYMENTS/CREDITS
JUN 3	FF	1390001	223	TELEVISION	311.50	
JUN 3	FF	1390011	99		3.47	
JUN 3		1390010	99			20.00
JUN 11		0670132		PAYMENT – THANK YOU		15.98

BILLING DATE	PREVIOUS BALANCE	PLUS: FINANCE CHARGE	PLUS: TOTAL PURCHASES	LESS: TOTAL PAYMENTS AND CREDITS	NEW BALANCE	PAYMENT FOR CURRENT BILLING CYCLE	29.00
JUN 14 87	15.98	0.00	314.97	35.98	294.97	PLUS PAST DUE AMOUNT	0.00
Periodic Rate		Applied to following portion of Previous Balance		ANNUAL PERCENTAGE RATE		EQUALS TOTAL MINIMUM PAYMENT DUE	29.00
1.6		0.00		19.2		DUE DATE	JULY 9

   NOTICE: See Reverse Side for Important Information. Purchases or credits not shown will appear on your next statement.

   To avoid FINANCE CHARGE on next statement, full payment of New Balance must be received by due date shown above.

   A. Would a purchase at Yorks on June 16 show on this bill? _____

   B. On what date did Yorks receive a $15.98 payment? _____

   C. If the bill is received on June 16, how many days does Joyce have to pay her bill in full to avoid any finance charge? (There are 30 days in June.) _____

4. Wes bought a motorcycle and made his first payment in July. He will need to make 14 payments in all. In what month should he make his last payment? _____

UNIT 8 ■ TIME   227

5. Charlene made a telephone call at 7:47 P.M. and talked until 8:35 P.M. How long was she on the telephone? _____

6. Mel started making dinner at 5:10 P.M. and the family sat down to eat at 6:55 P.M. How long did it take him to prepare dinner? _____

7. A $4\frac{3}{4}$-pound leg of lamb must cook 44 minutes a pound. If the roast is put in the oven at 3:15 P.M., at what time will it be cooked? _____

8. Melody needs to take medicine every 4 hours. She took some at 7:30 A.M. When should she take the medicine again? _____

9. If you plan to drive 200 miles to visit family and want to arrive at 7:30 P.M.,

   A. at what time must you leave, averaging 50 miles per hour? _____

   B. at what time must you leave, averaging 60 miles per hour? _____

10. Milly puts a cake in the oven at 10:10 A.M. The timer on the oven does not work. The cake takes 35 minutes to bake. When will the cake be done? _____

11. Dwight worked on his term project on March 13 and continued through March 21. How many days did he work on his project? _____

12. Leonor is planning to go out at 6:30 P.M. She has to shower (20 minutes), fix her hair (23 minutes), paint her nails (45 minutes), put on her makeup (16 minutes), and dress (14 minutes). At what time should she start to get ready? _____

TIME AT A JOB

On many jobs you will keep a record of the number of hours you work. The hours are recorded on a time card. It is important to write down the exact time you start work and the exact time you finish.

Example

Record the starting time, ending time, and time worked for Monday on the time card.

Monday

Start Work: 3:30

Finish Work: 5:45

| Week of _____ |
| Name _____ |
| Social Security # _____ |

Days	In	Out	In	Out	Regular Hours
Mon	3:30	5:45			2 hr 15 min
Tues					
Wed					
Thurs					
Fri					
Sat					
Sun					
Total Hours					

Answer the following questions.

1. Use the time card above to record the starting time, ending time, and time worked for each day below. Then record the total hours worked.

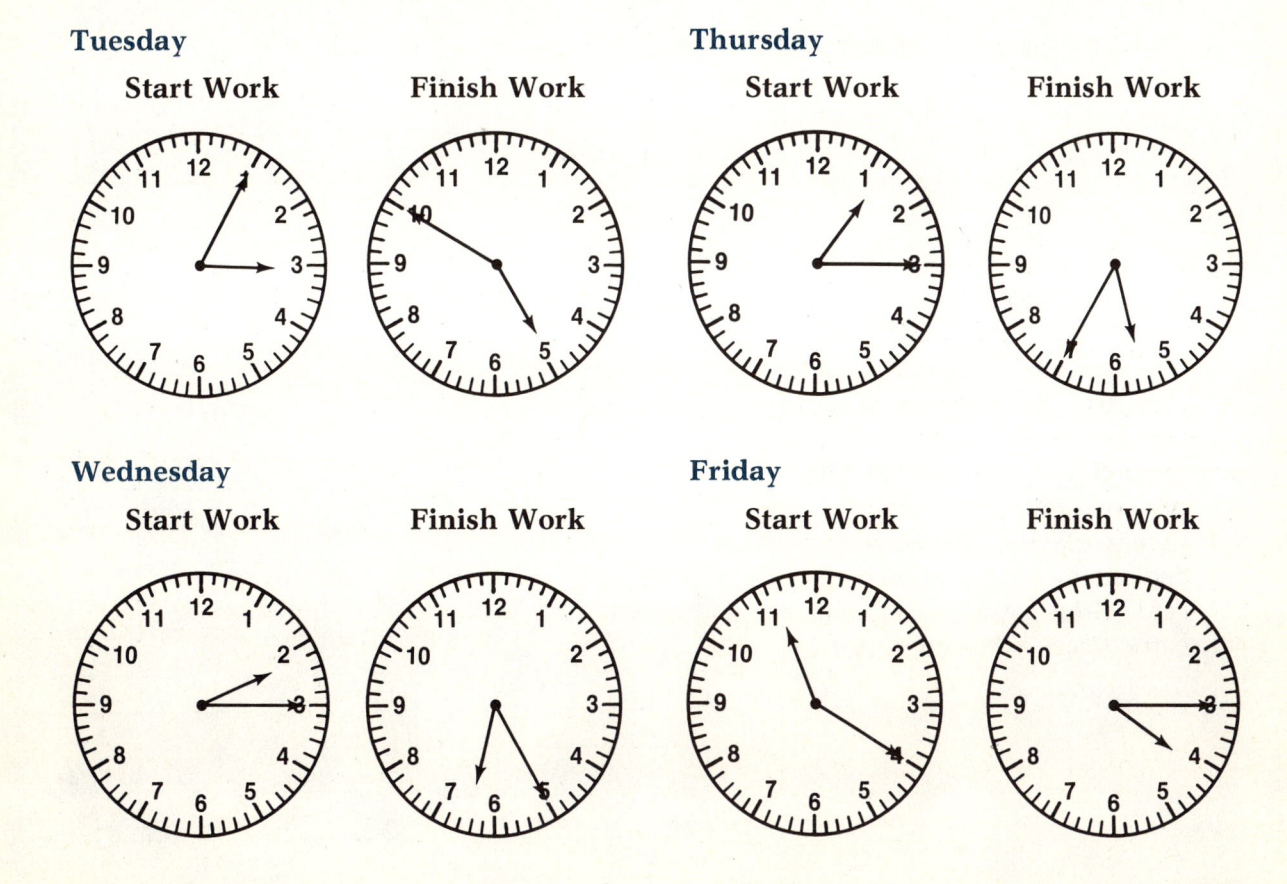

Tuesday

Start Work / Finish Work

Thursday

Start Work / Finish Work

Wednesday

Start Work / Finish Work

Friday

Start Work / Finish Work

UNIT 8 ■ TIME

229

2. Fill out the time card on the right for Lillian Shrieve. She worked Monday through Friday, from 8 A.M. until 5 P.M., with a lunch break from noon until 1 P.M.

Week of _____
Name _____
Social Security # _____

Days	In	Out	In	Out	Regular Hours
Mon					
Tues					
Wed					
Thurs					
Fri					
Sat					
Sun					
Total Hours					

3. Fill out the time card on the right for Lauren Henhart. She worked the second through the fifth day of the week from 7:30 A.M. to noon. Each day she took a lunch break from noon until 12:45 P.M., returned to work at 12:45 P.M., and went home at 4:15 P.M.

Week of _____
Name _____
Social Security # _____

Days	In	Out	In	Out	Regular Hours
1					
2					
3					
4					
5					
6					
7					
Total Hours					

4. Fill out the time card on the right for David Powell. He worked from 6 A.M. to 11 A.M., took a 1-hour lunch break, and worked from noon until 3 P.M. His regular pay rate is $5.25 an hour, and he worked Tuesday through Saturday. In addition to recording his time worked, figure his gross pay and enter the amount on the time card. (Gross pay is the total amount earned *before* deductions.)

Week of _____
Name _____
Social Security # _____

Days	In	Out	In	Out	Regular Hours
Mon					
Tues					
Wed					
Thurs					
Fri					
Sat					
Sun					
Regular Rate	×		Total Hours	=	Gross Pay
	×			=	

Computing Time and a Half

Overtime is the number of hours worked in addition to regular hours. The usual rate of pay for overtime is *time and a half*. However, some employees receive their normal rate of pay for working overtime, and some receive double time or an even higher rate of overtime pay.

The rate of pay for time and a half is $1\frac{1}{2}$ (1.5) times the regular rate of pay per hour.

Answer the following questions.

1. Complete the following table to figure total time-and-a-half overtime pay. Multiply the regular rate of pay by 1.5, and then multiply this answer by the number of overtime hours worked. The first one has been completed for you.

	Regular Rate Of Pay	×	Time and a Half	=	Overtime Rate	×	Overtime Hours Worked	=	Total Overtime Pay
A.	Bert $6.00	×	1.5	=	$9.00	×	3	=	$27.00
B.	Luisa $7.20						$4\frac{1}{2}$		
C.	Yoko $8.12						6		
D.	Ian $7.80						10		

2. Fill out the time card on the right for Deena Williams, including regular hours, overtime hours, and gross pay. She worked Monday through Friday. Each day she worked from 11 A.M. to 4 P.M., took a dinner break from 4 P.M. to 4:30 P.M., returned to work at 4:30 P.M., and worked until 8:00 P.M. If Deena works more than 8 hours in one day, the additional time is considered overtime and she receives time-and-a-half pay for those hours. Her regular rate of pay is $7 an hour.

Week of _____

Name _____

Social Security # _____

Days	In	Out	In	Out	Regular Hours	Overtime Hours
Mon						
Tues						
Wed						
Thurs						
Fri						
Sat						
Sun						
Total Hours						

Regular Hours × Regular Rate =

Overtime Hours × Time-and-a-Half Rate =

Total Gross Pay =

Figuring an Hourly Percent Raise

Some employers determine an hourly raise by using a fixed percent. To figure an hourly percent raise, change the percent to a decimal and multiply by the present hourly rate. Add this answer to the present pay rate to find the new hourly rate.

Answer the following questions. Round to the next higher cent as necessary.

1. Find the new hourly pay rate for the following employees and fill in the table below. The first one has been completed for you.

Name	Present Hourly Pay Rate	+	6% Increase	=	New Hourly Pay Rate
Wanda	$4.50	+	0.27	=	$4.77
Owen	$5.20				
Joy	$4.80				
Lonnie	$5.30				
Hugh	$6.40				

2. Using the table in Problem 1, if Wanda works 32 hours a week, how much more per week will she earn after her raise?

3. Use the table in Problem 1 to answer the following questions.
   A. If Hugh works 40 hours a week, how much will he earn in 4 weeks at his present pay rate?

   B. How much will Hugh earn in 4 weeks at his new pay rate?

4. Carlos and Darryl painted the inside of their neighbor's house. Carlos worked 12 hours and earned $78.
   A. At the same rate of pay, how much did Darryl earn for working 18 hours?

   B. If Carlos received an 8% raise, how much more per hour would he earn?

   C. At this higher rate of pay, how much would Carlos earn if he worked 15 hours?

UNIT 8 ■ TIME

USING TRANSPORTATION SCHEDULES

Reading transportation schedules becomes easy with practice.

| MONDAY THROUGH FRIDAY BUS SCHEDULE ||||||||||
| OUTBOUND ||||| INBOUND |||||
Leave Concord Station	Arrive Civic Center Salvio & Parkside	Arrive Willow Pass Rd. & Olivera Road	Arrive Willow Pass Rd. & Landana	Arrive Concord Blvd. & Clayton Way	Leave Willow Pass Rd. & Landana	Leave Concord Blvd. & Clayton Way	Leave Willow Pass Rd. & Olivera Road	Arrive Civic Center Salvio & Parkside	Arrive Concord Station
5:50	5:52*	5:55*	6:00	6:08	6:00	6:08	6:10*	6:12*	6:16#
6:50	6:52*	6:55*	7:01	7:09	7:01	7:09	7:11*	7:13*	7:17#
7:50	7:52*	7:55*	8:01	8:09	8:01	8:09	8:11*	8:13*	8:17#
8:55	8:57*	9:00*	9:06	9:14	9:06	9:14	9:16*	9:18*	9:22#
9:55	9:57*	10:00*	10:06	10:14	10:06	10:14	10:16*	10:18*	10:22#
10:55	10:57*	11:00*	11:06	11:14	11:06	11:14	11:16*	11:18*	11:22#
11:55	11:57*	12:00*	12:06	12:14	12:06	12:14	12:16*	12:18*	12:22#
12:55	**12:57***	**1:00***	**1:06**	**1:14**	**1:06**	**1:14**	**1:16***	**1:18***	**1:22#**
1:55	**1:57***	**2:00***	**2:06**	**2:14**	**2:06**	**2:14**	**2:16***	**2:18***	**2:22#**
2:55	**2:57***	**3:00***	**3:06**	**3:14**	**3:06**	**3:14**	**3:16***	**3:18***	**3:22#**
3:53	**3:55***	**3:58***	**4:04**	**4:12**	**4:04**	**4:12**	**4:14***	**4:16***	**4:20#**
4:57	**4:59***	**5:02***	**5:08**	**5:16**	**5:08**	**5:16**	**5:18***	**5:20***	**5:24#**
5:31	**5:33***	**5:36***	**5:41**	**5:49**	**5:41**	**5:49**	**5:51***	**5:53***	**5:57#**
5:57	**5:59***	**6:02***	**6:08**	**6:16**	**6:08**	**6:16**	**6:18***	**6:20***	**6:24#**
6:27	**6:29***	**6:32***	**6:38**	**6:46**	**6:38**	**6:46**	**6:48***	**6:50***	**6:54**
6:50	**6:52***	**6:55***	**7:00***	**7:08***	**7:00**	**7:08**	**7:10***	**7:12***	**7:16#**

Light Face Figures A.M.
Dark Face Figures P.M.
*Approximate Leaving or Arrival Time.
#Bus continues as Line 304 to Solano Way and Olivera Road. See Line 304 Timetable.

Use the schedule above to answer the following questions.

1. A. If you leave Concord Station at 4:57 P.M., what time will you arrive at Concord Blvd. and Clayton Way? _____

   B. How many minutes will the trip take? _____

2. It is 6:50 A.M. and Mr. Kalberer, who lives on Landana near Willow Pass Road, wants to catch the next bus to Concord Station.
   A. How many minutes does Mr. Kalberer have before the bus arrives? _____

   B. How long will it take to get to the station once he gets the bus? _____

3. Jim Taylor takes the 11:55 A.M. bus from Concord Station to his home on Olivera Road. How long will it take him to get to his stop on Olivera Road? _____

4. Are there more scheduled buses before or after noon? _____

5. What is the earliest time you could arrive at Concord Station by riding this bus? _____

UNIT 8 ■ TIME

Use the tables at the right to answer the following questions.

1. How many minutes is the bus ride from the Concord station to the Airport Plaza?

   _____

2. How many minutes is the bus ride from the Airport Plaza to the Concord station?

   _____

3. What is the earliest you could arrive at the Airport Plaza if you rode the bus from Concord?

   _____

4. What is the latest you could leave the Airport Plaza to ride a bus to Concord?

   _____

5. Your plane leaves the airport at 9:00 A.M. and you need to allow 30 minutes to check your luggage. What time should you catch the bus at Concord?

   _____

6. What might you do if you wanted to arrive at the Airport Plaza at noon?

   _____
   _____

7. A person riding a bus from Quincy to Reno, NV would arrive at Reno at what time?

   _____

8. A person riding the bus from Oroville, CA to Quincy would leave Oroville at what time?

   _____

9. How long would the bus ride from Oroville, CA to Sacramento, CA take?

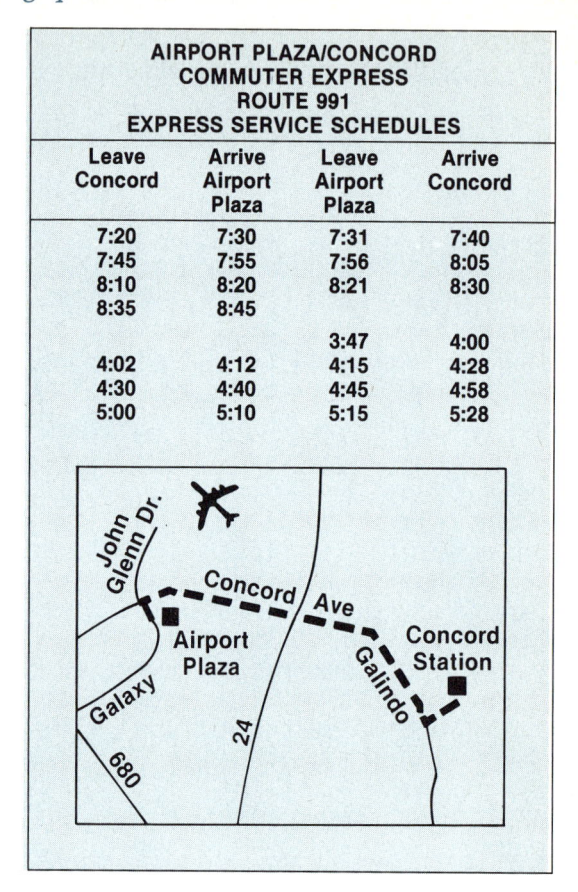

   _____

AIRPORT PLAZA/CONCORD COMMUTER EXPRESS ROUTE 991 EXPRESS SERVICE SCHEDULES

Leave Concord	Arrive Airport Plaza	Leave Airport Plaza	Arrive Concord
7:20	7:30	7:31	7:40
7:45	7:55	7:56	8:05
8:10	8:20	8:21	8:30
8:35	8:45		
		3:47	4:00
4:02	4:12	4:15	4:28
4:30	4:40	4:45	4:58
5:00	5:10	5:15	5:28

BUS TIMETABLE RENO—OROVILLE
Feather River Route

READ DOWN		READ UP
5997	SCHEDULE NUMBER	5995
9:05	Lv RENO, NV .... Ar	6:50
9:15	Harrah's (Reno)	6:40
flag	Jct. 70 & 395, CA	flag
flag	Chilcoot	flag
flag	Vinton	flag
flag	Beckwourth	flag
10:15	Portola	5:40
flag	Blairsden Jct	flag
flag	Feather River Prep School ..	flag
flag	Cromberg	flag
flag	Sloat Jct	flag
flag	Spring Garden	flag
11:00	Ar Quincy Lv	4:55
11:15	Lv Quincy .... Ar	4:40
flag	Feather River Jr College	flag
flag	Kaddie Jct	flag
flag	Paxton	flag
flag	Twain	flag
flag	Belden	flag
flag	Tobin	flag
flag	Storrie	flag
flag	Pulga Bridge	flag
1:10	Ar OROVILLE, CA Lv	2:45
1:15	Lv Oroville, CA .... (600) Ar	2:40
1:55	Ar Marysville, CA Lv	1:50
3:25	Ar Sacramento, CA Lv	12:25
2:55	Lv Oroville, CA .... (605) Av	1:10
3:45	Ar Chico, CA Lv	12:15
5:30	Ar Redding, CA Lv	10:30

USING TIME ZONES

There are 24 time zones. The continental United States is divided by 4 of these time zones. The 4 time zones are Eastern Standard, Central Standard, Mountain Standard, and Pacific Standard.

Time zones are determined so that the sun reaches its peak at around noon local time whether you are in New York, California, or Australia.

Study the map below and use it to answer the questions that follow.

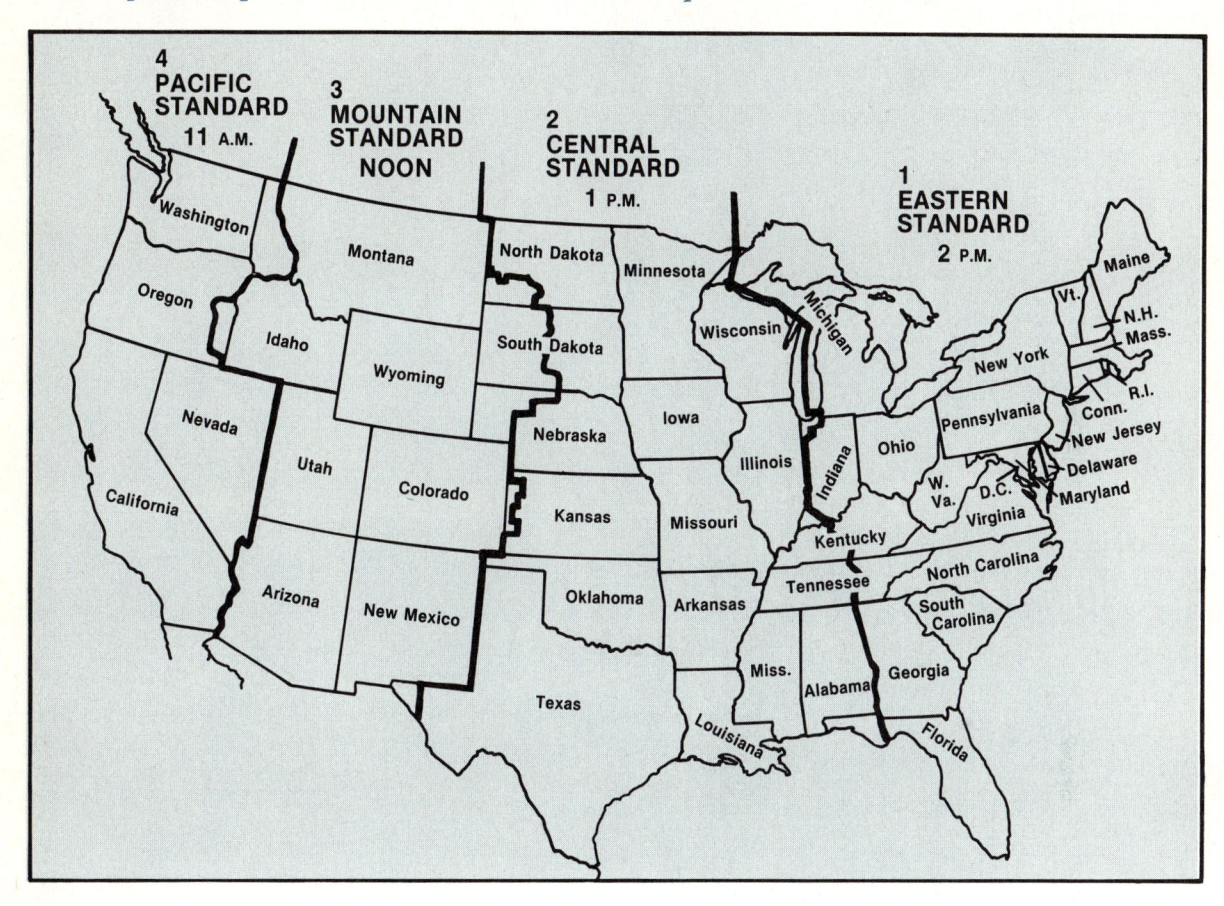

1. **A.** If you fly from Ohio to New Mexico, will you set your watch forward or backward when you arrive in New Mexico? _____

   B. If you leave Ohio at 5 P.M., to what time should you change your watch if you are traveling to New Mexico? _____

2. When it is 1 P.M. in Idaho, what time is it in each of the following?

   A. Texas _____   B. Nevada _____   C. Georgia _____

3. When it is 7 A.M. in Montana, what time is it in each of the following?

   A. West Virginia _____   B. Oregon _____   C. Arkansas _____

4. How long would a plane be in the air if it left Minnesota at 7:30 P.M. local time and arrived in California at 8 P.M. local time? _____

INDEX

A

addition
  of decimals, 13-14
  of fractions, 29-32
  of mixed numbers, 31
  of units of linear measurement, 159
adjusted balance, 120
area, 211-217
  of circles, 216-217
  of rectangles and squares, 212-216
automatic teller machines (ATM), 194-195
  receipt from, 195
  services at, 194
  steps for using, 194
automobile expenses, 116-117
average daily balance, 120
averages, 148-152
  mean, 148, 149
  median, 148
  mode, 148

B

balancing bank accounts, 196-198
  steps for, 196
bank accounts, 180
  balancing, 196-198
banking, 179-198
  automatic teller machines, 194-195
  bank statements, 192-193
  checking your balance, 196-198
  deposit slips and checks, 181-190
  filling out a check register, 190-192
  opening a bank account, 180
bank reconciliation worksheet, 196, 197
bank statements, 192-193
  identifying parts of, 193
  using to balance bank accounts, 196-198
bar graphs, 130
  horizontal, 130
  making, 139
  reading, 130-132
  vertical, 130
borrowing, 32-33
bus schedules, 125, 233-234

C

calculator
  clearing entries on, 2
  definitions of keys, 1
  estimating using, 72
  memory keys on, 3
  multiplying with equals key on, 2
  rounding with, 12
  using, 2-3
  using, to find cost per serving, 111, 112
  using, to find unit prices, 106
  using, to solve percent problems, 51, 53, 55
calorie tables, 124, 126
canceling, 24-25
catalogs, ordering from, 147
Celsius scale, 175
centi, 165
checking account, 180
check register, 190-192
  and automatic teller machines, 195
  filling out, 191
  using to balance bank account, 196-197
checks
  endorsing, 181
  writing, 186
circle, 201
  area of, 216-217
  center of, 201
  circumference of, 208-211
  diameter of, 208
  radius of, 208
circle graphs, 133
  estimating fractional parts of, 141
  making, 141
  reading, 133-135
circumference, 208-211
clearing entries on calculators, 2
coins, 182
combination prices, 102
comparison shopping, 99-121
  for automobile expenses, 116-117
  for credit, 118-121
  for groceries, 100-113
  for miscellaneous purchases, 113-115
cone, 202

convenience products, 109-110
cooking measurement, 171-174
  equipment for, 172
  equivalent units in, 171, 174
  increasing and decreasing recipes, 173
cost per serving, 111-113
  using calculator to find, 111, 112
  when using recipe, 112
coupons, 108
credit, 118-121
  annual percentage rate, 118
  cost of, 118
  department store, 120
cross multiplying, 42, 43
cube, 201
cups, 171
  equipment for measuring, 172
currency, 182
cylinder, 202

D

days, 225
  number of, from one date to another, 226
decimals
  adding and subtracting, 13-14
  changing from percents to, 37
  changing, to fractions, 39
  changing, to percents, 36
  dividing, 16-19
  metric units written as, 166-167
  multiplying, 14-15
  place value of, 4-9
  rounding, 10-11
  understanding multiplying and dividing, 19-20
decreasing recipes, 173
denominator, 20
  lowest common, 30
deposit slips, 182-185
  filling in, 182
diameter, 208
dividend, 16
division
  of decimals, 16-19
  in decreasing recipes, 173
  of fractions, 26-27
  of mixed numbers, 26
divisor, 16

E

ellipse, 201
endorsing checks, 181
equals key on calculators, 2
equivalent fractions, 29-30
equivalent units
  in area measurement, 211
  in cooking measurement, 171

equivalent units (continued)
  for figuring time intervals, 222
  of food, 174
  in linear measurement, 158
estimating, 70
  fractional parts of circle graphs, 141
  unit prices, 107
  when using calculator, 72

F

Fahrenheit scale, 175
federal income tax table, 144
feet, 158
formulas
  area of circle, 216
  area of rectangle or square, 212
  circumference, 208
  volume of rectangular object, 218
fractions, 20
  adding and subtracting, 29-32
  borrowing when subtracting, 32-33
  canceling when multiplying, 24-25
  changing from decimals to, 39
  changing from percents to, 37
  changing improper, to mixed numbers, 22
  changing mixed numbers to improper, 23
  changing, to percents, 38
  comparing size of, 20
  dividing, 26-27
  equivalent, 29-30
  improper, 20
  of inches, 154, 156
  multiplying, 24
  proper, 20
  reducing, 21
  understanding multiplying and dividing, 28-29

G

gallons, 171
geometric figures, 200-203
  three-dimensional, 201
  two-dimensional, 200
geometric shapes and calculations, 199-220
  area, 211-217
  geometric figures, 200-203
  perimeter, 204-211
  volume, 218-220
grams, 163, 164
  measurement using, 168-169
graphs
  bar, 130-132, 139-140
  circle, 133-135, 141-142
  line, 126-129, 137-138
  making, 137-143
  picture, 135-136, 143
  reading, 126-136

grid, 93
grocery shopping, 100-113
  convenience products, 109-110
  cost per serving, 111-113
  unit pricing, 100-108

H

half gallons, 171
horizontal bar graphs, 130
hourly percent raise, 232
hours, 222
  figuring parts of, 223

I

improper fractions, 20
  changing mixed numbers to, 23
  changing, to mixed numbers, 22
inches, 154, 158
income tax table, 144
increasing recipes, 173

K

keys of calculators, 1
  equals, 2
  memory, 3
  percent, 51, 55
kilo, 165

L

linear measurement, 154-162
  abbreviations for units of, 158
  adding and subtracting, 159
  equivalent units of, 158
  fractions of inches, 154, 156
  using a ruler, 155-157
line graphs, 126
  horizontal information on, 126-127
  making, 137
  reading, 126-129
  vertical information on, 126-127
liters, 163, 164

M

making graphs, 137-143
mean, 148, 149
measurement, 153-178
  cooking, 171-174
  linear, 154-162
  metric, 163-170
  temperature, 175-178
measuring equipment for cooking, 172
median, 148
memory keys on calculators, 3

meters, 163, 164
  measurement using, 167-168
metric measurement, 163-170
  comparing to U.S., 164
  prefixes in, 165
  units of, written as decimals, 166-167
  using grams, 168-169
  using meters, 167-168
metric prefixes, 165
metric units written as decimals, 166-167
milli, 165
minutes, 222
mixed numbers, 20
  adding and subtracting, 31, 32-33
  changing improper fractions to, 22
  changing, to improper fractions, 23
  dividing, 26
  multiplying, 24
mode, 148
months, 225
multiplication
  of decimals, 14-15, 19
  of fractions, 24
  in increasing recipes, 173
  of mixed numbers, 24

N

numerator, 20

O

opening a bank account, 180
oval, 201
overtime, 231

P

parallelogram, 200
pattern table, 146
percent key on calculators, 51, 55
percent problems, 46-58
  finding a number when a percent of the number is known, 54, 55
  finding percent of a number, 50, 51
  finding what percent one number is of another, 52, 53
  other approaches to solving, 50-51, 52, 53, 54, 55
  solving using cross multiplying, 46-47
percents, 36, 46-50
  changing from decimals to, 36
  changing from fractions to, 38
  changing, to decimals, 37
  changing, to fractions, 37
  raises as, 232
  solving problems with, 46-58
  with fractions, 40

perimeter, 204-211
  of rectangles, 204, 205
pi (π), 208
picture graphs, 135
  making, 143
  partial pictures in, 136
  reading, 135-136
pints, 171
place value, 4-9
  visual explanation of decimal, 7
place value chart, 4
polygon, 202
previous balance, 120
prices
  combination, 102
  unit, 100-108
problem solving strategies, 60
  decide which math operation to use, 68
  draw a picture or diagram, 87
  estimate, 70, 72
  follow directions, 60
  guess and check, 85
  keep track of clues and information, 92
  look for likenesses and patterns, 80
  manipulate objects, 89
  read problem carefully and look for key words, 64
  simplify and substitute easier numbers, 74-75
proper fractions, 20
proportion, 42
  finding missing number in, 43-44
pyramid, 202

Q

quarts, 171

R

radius, 208
ratio, 41
reading graphs, 126-136
reading tables, 124-126
reconciliation worksheet, 196, 197
rectangle, 200
  area of, 212
  perimeter of, 204, 205
reducing fractions, 21
right angle, 200
rounding decimals, 10-11
rounding money, 11-12
  store method of, 102
rounding numbers, 9-12
rounding whole numbers, 10
rounding with calculators, 12
rulers, 154
  fractions of inches on, 154, 156
  measuring using, 155-157

S

safe deposit box, 179
sales tax table, 124
savings account, 180
signature card, 180
sphere, 201
square, 200
  area of, 212
strategies for solving word problems, 60
subtraction
  of decimals, 13-14
  of fractions, 29-33
  of mixed numbers, 31, 32-33
  of units of linear measurement, 159

T

tables
  bus schedules, 125, 233-234
  calorie, 124, 126
  catalog, 147
  for converting units of time, 222
  federal income tax, 144
  pattern, 146
  reading, 124
  sales tax, 124
tablespoons, 171
  equipment for measuring, 172
tally, 93
teaspoons, 171
  equipment for measuring, 172
temperature measurement, 175-178
thermometer, 175
three-dimensional figures, 201
time, 221-235
  figuring time intervals, 222-228
  time at a job, 229-232
  using time zones, 235
  using transportation schedules, 233-234
time and a half, 231
time cards, 229, 231
time intervals, 222-228
  conversion table for, 222
  hours and minutes, 222-224
  months and days, 225-228
time zones, 235
triangle, 200
two-dimensional figures, 200

U

unit prices, 100-108
  changing dissimilar units before finding, 106
  estimating, 107
  finding, 101, 102
  using calculator to find, 106
  when using coupons, 108

U.S. measurement, 163
   comparing to metric, 164

V

Venn diagram, 96
vertical bar graphs, 130
volume, 218-220
   of rectangular object, 218

W

whole numbers
   place value of, 4-9
   rounding, 10
word problems and problem solving, 59-98
   strategies, 60
writing checks, 186-190

Y

yards, 158